Praxiology and the Philosophy of Technology

Praxiology and the Philosophy of Technology

Praxiology:
The International Annual of
Practical Philosophy and
Methodology

15

Edited by

Wojciech W. Gasparski
Timo Airaksinen

Routledge
Taylor & Francis Group
NEW YORK AND LONDON

First published 2008 by Transaction Publishers

Published 2021 by Routledge
605 Third Avenue, New York, NY 10017
2 Park Square, Milton Park, Abingdon, Oxon OX14 4RN

Routledge is an imprint of the Taylor & Francis Group, an informa business

Library of Congress Catalog Number: 2007029486

Library of Congress Cataloging-in-Publication Data

Praxiology and the philosophy of technology / [edited by] Wojciech W. Gasparski and Timo Airaksinen.
p. cm. — (The international annual of practical philosophy and methodology ; v. 15)
Includes bibliographical references.
ISBN 978-1-4128-0682-4 (alk. paper)
1. Technology—Philosophy. 2. Technology—Moral and ethical aspects. 3. Praxeology. I. Gasparski, Wojciech. II. Airaksinen, Timo, 1947-

T14.P73 2007
601—dc22

2007029486

ISBN 13: 978-1-4128-0682-4 (hbk)

Contents

Editorial

Wojciech W. Gasparski
Editor-in-Chief

Praxiology and Philosophy of Technology are twin disciplines. They were born as successors of the same parents devoted to practical thinking and doing. Those who look for their pre-predecessors usually refer to Aristotle and his immediate followers. Let us leave the archaeology of the disciplines to others, limiting ourselves to the recent history of praxiology and its twin sister, philosophy of technology.

Initially, praxiology was the *science of functions* situated in Louis Bourdeau's system of sciences, next it was the most general *science of techniques* characteristic for human action approached by Alfred Victor Espinas, then it was considered by Tadeusz Kotarbinski as a *general methodology*, and by Ludwig von Mises as a theoretical *foundation of economics*, finally Mario Bunge situated it within the family of philosophical disciplines dealing with ethics *sensu largo* or practical philosophy as *human action theory*, while Jan Zieleniewski situated it among *ergological sciences*, i.e. studies on human labor. In addition to the above, one may conclude that praxiology, located between logic and ethics, offers epistemological foundations for the concept of *reflective practitioner*, pointing out the *axiological context* of human action. Let us revisit the story of praxiological inquiry and its sisterhood with philosophy of technology.

The French scholar Louis Bourdeau (1824–1900) published his *Théorie des sciences: Plan de science intégrale* in 1882. In this treatise he criticized the fact that science of that era was unable to classify rationally the problems studied by particular disciplines (*sciences particuliées*). He suggested that the disciplines should be replaced by general sciences (*sciences générales*) dealing with classes of phenomena; furthermore, the superior science (*science intégrale* or *science des sciences*) should offer methodological tools enabling a transition from setting collections of entities a priori to aposterioric definitions of classes of phenomena. To be considered a science – according to Bourdeau – an approach has to fulfill three conditions: to focus on a well-defined subject of studies, a research program, and a method. Praxiology as a science of functions was situated among other general sciences dealing with the following classes of facts: existence (ontology or logic), quantity (metrology or mathematics), collocation (teseology or dynamics), modality (poiology or physics), composition (craseology or chemistry), and structure (morphology).

Function, according to Bourdeau, is a series of results (effects) that occur in specific forms as outcomes of the environment's influence. The plastic individuality of living beings gives them the possibility to act by themselves. They detect the environment's influences and react to them, modifying it to their needs and existence. Every body whose form is stable and consolidated has functional features. The force of activity (similar to gravity in physics) is responsible for functions. The task of praxiology (πραξιξ = action) or "mesology" (μεσοξ = environment) should be, according to the above-mentioned author, to study issues related to this area. He considered practical sciences as its sub-disciplines. Was this a prediction of technology and/or engineering of a different kind, understood as "applied" science? Maybe; there is no evidence either for or against this supposition.

The situation was completely different with another French scholar, Alfred Victor Espinas (1844–1922), who published a paper (1890) and then a treatise (1897) under the same title, *Les origines de la technologie*. His fundamental theses presented in these works say that human abilities, like animal instincts, have two different important characteristics: instincts are forms of activity that are inherited by an organism, while abilities are products of experience and reflection preconditioned by invention, initiative and freedom. Every social group is defined by its abilities no less that every species by its instincts.

Useful abilities or *techniques* (τεχναι) are conscious and related to reflection of practical deeds. They are in opposition, to some extent, to primitive acts performed spontaneously without previous analysis. Mature abilities give birth to a science of which *Technology* is a part while

Praxiology is the most general part of that science.[1] Three types of studies: (i) analytical description of practical abilities (statics), (ii) research on conditions, laws, rules, causes that guarantee effectiveness (dynamics), (iii) research on evolution of techniques applied by people in their activity, create *general technology* or *praxiology*. Praxiology understood in such a way plays a similar role in the area of action as logic plays in the area of cognition, studying and classifying different disciplines in a similar manner, defining their conditions or laws, reconstructing their development or history. Once philosophy of cognition or epistemology had its historians, wrote Espinas, it would be useful to do the same in the area of philosophy of action. This is a task, one of the tasks, of praxiology pointed out by the French scholar and/or – let us add – of philosophy of technology.

Polish philosopher Tadeusz Kotarbinski (1886–1981), a student of Kazimierz Twardowski, the founding father of the famous Lvov–Warsaw school of philosophy, logic and mathematics, created a philosophical system which I have proposed to call Philosophy of Practicality.[2] At the core of this system is practical philosophy or ethics in the broader sense of the term, i.e. wisdom or theory of directing the spiritual life of human beings.

According to Kotarbinski, practical philosophy comprised: (i) felicitology (hedonistics, eudaimonology) or the study of a happy life; (ii) praxiology (general methodology, general theory of action) or the study of the practical dimension of human actions; (iii) ethics in a narrow sense (ethics proper, moral deontology) or the study on "how one should live to be called an honest person". The task of the studies is to construct the most rational programs of conduct from the point of view of satisfaction, efficacy and fairness. Kotarbinski considered practical philosophy understood in this way (ethics in a broader sense) as a practical discipline, and ethical culture as an element of philosophical culture.

Praxiology was given different names by Kotarbinski during the period of his involvement in its creation: *practice* (or general practice), *theory of deed*, *general methodology*, and *theory of efficient action*. These names reflect the different aspects of human action Kotarbinski was busy with when he developed the core component (subsystem) of the philosophy of practicality, a philosophical system he formulated. Praxiology also received other triggers to be developed by other authors according to the ancient saying: *habent sua fata libelli*, suggesting to its reflective creator new aspects worth considering.

The beginning of Kotarbinski's praxiology overlaps with the beginning of his philosophy. In 1910–1913, he analyzed concepts of the goal of an act and the task of the agent, position, compulsory situation and many others. Being influenced by Espinas' work and Bogdanov's *tectology*, he subsequently decided to name his studies *praxiology* (1923) and formulated the principles of a deed theory. At the time, he formulated an analytical definition of the concept of authorship of a deed, later modifying it to link it with the natural sequence of events. The draft of his praxiological treatise was destroyed by fire during the Warsaw Uprising (1944) and was never reconstructed in full. It served him as a basis for the first lectures he delivered after the war, during the period of his peregrination through the towns of war-ravaged Poland when he met a number of refugees hungry also for intellectual recovery. The *opus magnus* of Kotarbinski's praxiology, *The Treatise on Good Work* was published in 1955 in the completely different circumstances of postwar Poland. Its English translation was published in 1965 under a different title: *Praxiology: An Introduction to the Sciences of Efficient Action.*[3]

All the works on praxiology by Kotarbinski are collected in three large volumes of his *Collected Works* (published in Polish): *Praxiology, Part One* (1999, 462 pages) containing works published before the Second World War; *The Treatise on Good Work* (2000, 220 pages), and *Praxiology: Part Two* (2003, 699 pages) containing works published after the Second World War. No one else has made such a rich contribution to the study of praxiology.

A parallel to the Polish school of praxiology developed by Kotarbinski and his followers is the Austrian school initiated by Ludwig von Mises (1881–1973). He was an economist whose studies on the foundation of economics yielded the extensive treatise *Human Action: A Treatise on Economics* (1949). He pointed out that praxeology[4] is the basis on which economics is founded. The concepts of exchange and cooperation are essential here. The subject of praxeology, he wrote, and its best developed part, economics, is to study the third class of laws, i.e. laws of "the regularity of phenomena with regard to the interconnectedness of means and ends, viz. the praxeological law as distinct from physical and from the physiological law."[5] In relation to Mises' praxiological study one may pose the question: to what extent is economics a kind of technology or engineering?

Contemporary Canadian philosopher Mario Bunge is probably closer to considering economics as scientifico-technical, i.e. as a part of social science provided it would follow the Realistic Decalogue[6] or as technology (extended from engineering to decision theory) especially in relation to management science and social engineering.

Like Kotarbinski, Bunge situates praxiology within practical philosophy[7] or ethics, being about what is good and what is right. The difference between the two philosophers is in the system of which praxiology is a subsystem. For Bunge the system is composed of:

- a value system that is studied by axiology (theory of values), composed of scientific axiology, philosophical axiology, and meta-axiology or philosophy of concepts and theories of values;
- moral codes that are studied by ethics *sensu stricto* (theory of morals), composed of scientific ethics, philosophical ethics, and meta-ethics or philosophy of concepts and theories of morals;
- patterns of action that are studied by praxiology (theory of action), composed of scientifico-technical praxiology, philosophic praxiology, and meta-praxiology or philosophy of action concepts and theories[8]

There are also authors – like Polish management science scholars Jan Zieleniewski and Tadeusz Pszczolowski – who count praxiology among the "ergological sciences" studying human labor and other techniques applied in human actions. Others consider praxiology more broadly, as related not only to humans but to all living beings, suggesting *biopraxiology.*[9]

For many authors praxiology is a theoretical and/or methodological basis for other disciplines of technological character, e.g. of the *sciences of the artificial*, as Herbert A. Simon called applied disciplines.[10] It is necessary to stress that both Kotarbinski and Simon identify design as a characteristic feature of the practical/applied sciences. American technology philosopher Carl Mitcham suggests that the praxiological aspect is widely present in the practice of technology and in disciplines that serve as its theoretical basis.[11]

Praxiology today plays a similar role in the domain of action as logic plays in the domain of cognition.[12] However, logic is not enough. Efficacy of action depends not only on nature, but also, maybe to a greater degree even, on culture. Praxiology as such offers an insight into the reality it studies, but the use of its discoveries depends on people, as the actors. This use is more indirect than direct and needs the actor's reflection of his or her own practicality. Therefore the more an actor is a *reflective practitioner*,[13] the greater is the efficacy of his or her action.

A reflective practitioner is an actor who, according to Donald A. Schön,[14] has the following skills: (i) is able to acquire new abilities, (ii) is able to acquire new knowledge, (iii) is able to prepare an action conceptually (to design), (iv) is able to evaluate an action multidimensionally in the space I have called the "triple E" for effectiveness, efficiency and ethicality.[15]

This volume devoted to interrelations between praxiology and philosophy of technology is an attempt to facilitate the process of understanding human action in which unlimited numbers of tools produced intellectually and materially by technology are used as well as calling for better informed reflection on their design and use.

Notes

1. J.J. Ostrowski, *Alfred Espinas, précurseur de la praxéologie*, Paris 1973; V. Alexandre & W.W. Gasparski, eds., *The Roots of Praxiology: French Action Theory from Bourdeau and Espinas to Present Days*, Transaction 2000, New Brunswick (USA)
2. W. Gasparski, 1993, *Philosophy of Practicality: A Treatise on the Philosophy of Tadeusz Kotarbinski*, Societas Philosophica Fennica, Helsinki.
3. T. Kotarbinski, 1965, *Praxiology: An Introduction to the Sciences of Efficient Action*, Pergamon, Oxford.
4. Both Espinas and Mises spelt the name of the study as 'praxeology' many others spell it as 'praxiology'.
5. L. von Mises, 1966, *Human Action: A Treatise on* Economics, Contemporary Books, Inc., Chicago, p. 885.
6. M. Bunge, 1985, *Treatise on Basic Philosophy*, Vol. 7, *Epistemology & Methodology III: Philosophy of Science and Technology*, Part II, *Life Science, Social Science and Technology*, Reidel, Dordrecht, pp. 178–193.
7. *Practical philosophy* or ethics *sensu largo* should not be understood as the *philosophy of practicality*, i.e., a philosophical system of Tadeusz Kotarbiñski.
8. M. Bunge, 1989, *Treatise on Basic Philosophy*, Vol. 8, *Ethics: The Good and the Right*, Reidel, Dordrecht, p. 1.
9. A. Kaufmann, 1968, *The Science of Decision-making: An Introduction to Praxiology*, Weidenfeld & Nicolson, London.
10. H. A. Simon, 1981, *The Sciences of the Artificial*, 2nd Ed., The MIT Press, Cambridge, Mass.
11. Mitcham, C., 1994, Thinking through Technology: The Path between Engineering and Philosophy, The University of Chicago Press, Chicago.
12. W. Gasparski, 1996, Between Logic and Ethics: The Origin of Praxiology, *Axiomathes*, Vol. VII: 3, pp. 385–394.
13. D.A. Schön, 1992, The Crisis of Professional Knowledge and the Pursuit of an Epistemology of Practice, w: J.L. Auspitz, W.W. Gasparski i in. red., *Praxioologies*

and the Philosophy of Economics, Transaction Publishers, New Brunswick (USA)–London (UK), pp. 163–186.
14. D.A. Schön, 1983, *The Reflective Practitioner: How Professionals Think in Action*, Basic Books, New York.
15. W.W. Gasparski and L.V. Ryan, eds., 1996, *Human Action in Business: Praxiological and Ethical Dimensions*, Transaction Publishers, New Brunswick (USA)–London (UK).

Introduction

Timo Airaksinen
University of Helsinki
Finland

Technology is a crucial feature of the modern world. We should want it used effectively, efficiently, and ethically, or not at all. But, if we bother to reflect on the issue, we realize very quickly that the last alternative does not exist. Human beings are always and necessarily technological creatures. We have hands that are adept at using tools and intelligence that guides our actions. We also have our desires and emotions which motivate us to achieve more and more, never allowing us to stop. It often appears that we have no time to stop and reflect on our actions and their consequences. Humans may be brilliant at creating tools and technologies, and we may even be able to use them effectively and efficiently. But, we do not seem to care very much for the ethical foundations and consequences of our actions.

We need to stop and step outside the immediate context of work and achievement and watch what we have done and what we will do in the future. This is what the original meaning of "theory" used to be in Ancient Greece: watch the phenomena as they are. Nothing is more instructive or difficult than to adopt an impartial, detached view-point from which the past, present, and the future can be seen as they are. Perhaps this is impossible. The world is too complex and its past ambiguous, the present time complex, and the future genuinely open. Yet, it is our moral duty to try to be theoretical as well as efficient and effective. Only then is ethics possible.

We are responsible for the future generations who can expect that we will leave them a better world than we got. This is a crucial point because our modern world is technological through and through. In other words, it is man-made and not natural. Our modern world is artificial, invented, manufactured, and constructed. For this reason it looks like us, its makers in the great act of a second creation. Technology is like a mirror in which its creators and users can see their own face. Whether it is ugly or beautiful is our own decision. But, we should not deceive ourselves. We must be honest and willing to see things as they are, and if they are not good, we must fix them. Otherwise, our great technological inventiveness combines with blind effectiveness which is more destructive than creative. The world that we leave for the next generation must not be destroyed.

In the philosophy of technology there are three concepts we need to review before looking into the contents of the present volume in more detail. They are technological determinism, "the silence of technology," and neo-Luddism. The articles in this volume implicitly reject all three notions and, by doing so, reinforce our trust in a better technology in the future.

Technological determinism, as a doctrine, says that our social world and its values change according to its technological change. What such a thesis means is, of course, unclear and a subject of a rather vicious debate. But, its main tenet is clear: technology drives history and determines its outcome. Hegel's early example is taken from war and warfare. The virtues of manliness and courage have always been dear to the war faring classes and their subordinates. Men are manly, and that means courage when facing physical danger and the threat of death. The opposite is cowardice, which is worse than death. But, such a classical Roman notion of courage is dependent on the type of combat at that time. Weapons were designed for man against man duels and required strength, stamina, and aggression. In this non-causal sense, weapons determine the meaning of courage and manliness. Then, says Hegel, guns and rifles are invented, and the concept of courage changes accordingly. Now, soldiers fight without meeting each other on the battle field. They shoot at an enemy they can hardly see, and they die without ever facing a living adversary. They do not need aggression; on the contrary, they need to stay cool and calculating. They do not hit another person with a sword; they aim their rifle and pull the trigger. This is a kind of abstract and non-personal courage. Soldiers do not wear fancy red and blue uniforms, they wear camouflage. They become invisible to the enemy. Yet, even they need all the courage they can find.

Another example is mass tourism which could only flourish after large jet passenger airplanes became so common that the price of a trip was low enough. Now, all of the people must travel. They can no longer live without travel. Such examples are easy to coin. What they show is another thing.

The opposites of technological determinism is even more nebulous, namely voluntarism and constructivism. The first one says that technology serves the human need and thus we make innovations that are useful to us. In the 19th century industrial fires were a main problem. Large factories were constructed, but they were fire hazards. It became clear that fireproof buildings were badly needed, so engineers and architects started solving the relevant problems, successfully. Of course, nothing is perfect in the world of technology, as the surprise 9/11 collapse of the Twin Towers proved. Nevertheless, the main point is clear: we need fireproof buildings and this is what we get. The technology is created and the need satisfied. This is the essence of technological voluntarism.

Of course, most of the technological innovations are not originally desired, needed, or anticipated. A good example of an innovation which was anticipated but not needed was the airplane. When the Wright-brothers started flying, no one wanted their machine because they did not know what to do with it. The cadget worked well enough, and it flew long distances without crashing, but it was obviously useless. Yet, many people had tried to fly earlier. An example of an invention that was not anticipated and was useless is the PC. Nobody knew what to do with them, except use them as fancy typewriters. When the WWW came along, all that changed. Now, we know what we should do with them. If such examples carry any weight at all, then technological voluntarism cannot be the whole truth about the driving forces behind the modern world.

Praxiologically speaking, technology does not exist merely to satisfy the human need in an effective and efficient way. It is not possible to want and desire a goal that is met and satisfied simply by the means of "good work." The problems of the modern world and the questions we must ask are much more complicated than what the voluntaristic model dictates. Our preparatory actions are complicated, demanding, and initially mysterious, perhaps even partly subconscious.

Today many theorists would like to resort to the constructivist model of technological progress. The point is simple: technological inventions and innovations come and go, they are everywhere, but only some of them are actually adopted and put to use. What is done with them depends on the users and their imagination. The technologists can just offer them to the public and react to what happens. For instance, cellular phones are now used to send text messages and photographs. These functions are new and were originally and largely unanticipated until they were constructed by the users as the main functions of the new tool. The users are as creative as the engineers ever were. For this reason, we need to think of technology in a new way. Effective tools reach the goals we set for them, but what those goals are does not need to be clear when the tools are first launched. Technological means can be judged to be effective only after we know what they are for. Text messaging is an effective (and efficient) means of communication, as the relevant youth culture so clearly indicates. The praxiology of technology has many new problems to solve, and this is why the articles in this volume are needed.

The "silence of technology" is another problem that needs to be addressed in theory and in practice. Technology is silent in the sense that the public tends to forget about it. Notice how paradoxical this is. First, we say that users are technologically creative and construct the effective means of coping in the changing world. Next, we claim that technology is silent, that is, it passes unnoticed in the course of modern life. How do constructivism and the idea of silence fit together? This is one of the tough questions in the praxiology and philosophy of technology. People seem to experience technology somehow, but yet, they do not conceptualize this silent knowledge. The result is a tendency towards an irrationalism about technology and its uses.

In any case, we rarely experience the technological world around us. Technology is like white noise that we only notice when it stops, or in the case of technology when our tools and means stop working. As long as they work as they should, or as we have learned to expect, we do not pay attention. How can we create efficient and effective means if we do not pay attention? We may fix problems, but that is not enough when large systems are wasteful, costly, unreliable, irrelevant, and out of date. We let them stay in place as long as nothing alarming happens. We do so simply because we do not bother to pay attention. We do not realize that the water we drink and the air we breath are products of various technologies. In an urban environment hardly anything is "natural" these days. Wood is actually plastic; clothes are made of microfibers, and wine – who knows how it is made in those modern industrial wineries which produce millions of bottles per year. We live inside a giant machine made of technological products, their constellations, and interconnected systems. Such a total environment is all too easy to dismiss and forget. Yet, we insist that it should work well and satisfy the stringent norms we ideally set for it. It may all be wishful thinking and self-deception, but in order to see what it is, we must break the silence of technology first. This is the reason why we need to examine technology – and to talk about it – in a way which is independent of the engineering work. As Martin Heidegger said, the essence of technology is not something technological. This book is an attempt in that direction.

When people, the users of technology, finally hear the call of technology, they are often horrified. They think it is all artificial, dangerous, anti-traditional, and polluting. They do not see its ethics. They reject all the claims to efficiency and effectiveness. They panic and want to get rid of technology altogether. But, now it is too late. We are living in a technological world that cannot be dismantled, not without a large scale technological effort, or it will endanger our well-being and even our survival on the Earth. People become technological pessimists and neo-Luddites when they finally recognize what is happening around them.

As such, technology is an optimistic project based on one of the original master narratives of the modern era: Progress. Technology is the motor that moves progress, that is understood as a human good and universal right. We all profit from it, and we can claim its fruits. Historical evidence may support such an optimistic view. Food production, medical care, housing, transportation, and communication are all examples of technological success stories. We now live much better lives than our ancestors ever did, therefore we should not complain.

However, the dark side exists as well, as the post-modern techno-pessimists and neo-Luddites emphasize. War and violence are widespread, poverty is a problem, and most of all, the destruction of nature looks imminent, especially if we focus on global warming. More subtle problems exist as well: modern life is intellectually and emotionally shallow and superficial. Our activities are uninteresting and devoid of human value. Mass society and commercialism are constant threats. Genuine human relationships are threatened by the mad messages of the mass media. The list goes on for ever showing us how well-founded techno-pessimism is.

For this reason, ethical considerations are imminent. Can we trust the producers of technology? Can we trust the users? When technology is commercialized, how are the relevant responsibilities created, distributed, and controlled? Are the free markets responsible at all, or is this notion already obsolete? Technology and the market are inseparable now, after the Soviet Union failed in its attempt to utilize technology on a large scale without allowing the market forces to shape it. As we know, this effort was a complete failure on all fronts. Millions of people suffered from this experiment, therefore we should have learned the lesson that markets are needed. At the same time responsible behaviour is more important than ever. Neo-Luddism has a point as long as technology and its commercial applications are not under responsible control. The free world is free in the sense that excludes responsibility and its ethics. Can we trust technological players?

In this book the authors address the important question in the praxiology, philosophy, and ethics of technology by trying to point out the road not only to its better applications but to its better understanding. Heraclitus says that unexamined life is not worth living. In the same way one can claim that unexamined technology is not worth applying. It creates more harm than benefits. We have a moral duty to examine our tools and not just assume, naively, that they are neutral and innocent simply because they are tools.

The first part of the book, the papers by Tondl, Gasparski, and Herfel, address some general questions about the theory of technology. Tondl asks in his magisterial contribution the basic question of the technological world; how should it be understood, and by doing so sets the tenet of the book. Gasparski does the same with praxiological notions in the spirit of the great Tadeusz Kotarbinski. Herfel provides us with a picture of technological progress and the rather surprising processes which guide it like an invisible hand. He pays attention to path dependence in action. Praxiology has much to say about action.

The second part of the volume focuses on some special topics like ICT (information and communication technologies (Klincewicz) and work systems (Bedny and Karwowski). Next, we look at the foundations of problem solving (Nadler and Chandon) and the history of innovation in the age of enlightenment (Ranea). The unifying theme here is praxiological effectiveness: how to create and design systems which actually do what they are supposed to do, that is, which are effective tools and how to make them desirable?

The third part of the book is ethical. The main point is that ethics is needed when we apply technology and, especially, when we commercialize it in order to meet the needs of the people in a free market society. Airaksinen and Paukkonen start by asking when we can call technology human-centred. They make it clear that it need not be so. Heikkerö discusses Alfred Borgman on technological culture. Borgman is one of the leading ethicists of the technological world. Takala discusses an all-important field of applied technology, medicine. Any techno-optimist needs to mention this field, just like any pessimist must discuss the weapons industry. The last three papers move on from pure theory to the field of corporate life in a technological environment. Ozarek discusses ICT and the important praxiological notion of quality. Allinson and Lewicka-Strzalecka both deal with corporate responsibility. Individual ethics and responsibilities of a corporation in a market situation saturated with technology and its demands is really a moot question.

This book should offer its reader a balanced picture of the state of the art of the theoretical and applied issues in technology studies, emphasizing the praxiological perspective.

Acknowledgements

Takes Finland has supported Timo Airaksinen's research.

PART ONE

Technology and Praxiology

Problem Areas of the Philosophy of Technology

Ladislav Tondl
Institute of Philosophy,
Academy of Sciences
Prague
Czech Republic

1. Essential Features of the Technological World

When asked to specify more precisely the actual nature and function of the environment in which contemporary man lives, creates and accomplishes most of his goals, intentions and lifelong ambitions, we come to realize that this environment is composed – to a large extent – of objects, processes and conditions which have been created by man himself and which we sometimes collectively call the technological world. (Many languages and cultures prefer using the attribute "technological" consisting of two original Greek terms, i.e. "techne" expressing man's capacity, skills and also the art of coping with or solving a given problem situation or reaching a defined goal, and "logos", a term denoting not only word but also an idea, knowledge or learning. Seen in this light, when required to translate this particular attribute in the Anglo-Saxon but also French-speaking countries adequately, we should rather use the term "technological".)

The technological world can hardly be characterized or simply defined by a single criterion. The main difficulty also lies in that this particular world or, as some thinkers say, this "other nature" is not homogeneous, encompassing a multitude of considerably incongruous units, namely those of ideate, energy as well as material nature, which are intertwined in a highly variegated fashion. With this in mind, we shall attempt here to outline at least the most significant characteristics and criteria.

1. The technological world is primarily something that may be associated with the term "*artificial*", i.e. something that is created by a goal-directed human activity, based on knowledge, skills or ascertained and verified experience. The attribute "artificial", as a substantial trait of the technological world, has already been emphasized by H. Simon, one of the leading pioneers of the so-called artificial intelligence.

2. The technological world creates and also partly supplements or replaces the current nature around us by what is sometimes referred to as the "*second nature*". Having said that, we also mean that the technological world in no way changes the rules or laws disclosed by the natural sciences, on the contrary it must fully respect them. To put it in other words, the "second nature" is subject to the same principles and rules as those disciplines studied by contemporary natural sciences. After all, it was F. Bacon who spelt out the well-known principle that nature may be utilized and controlled only if we respect its laws (Natura nil nisi parendo vincitur.)

3. A substantial part of elements and components of the technological world may be viewed as different forms of utilizing or *applying knowledge*. (Due to these reasons, one of the current trends in the philosophy of technology, whose leading representative is primarily M. Bunge, regards technological and engineering branches as applied natural sciences.) Engaging one's knowledge is, undoubtedly, an essential constituent of all the major spheres of technological activities covering both intellectual and information preparation of technological solutions as well as the implementation and utilization of all the forms of the results of technological activities. At the same time, it should be emphasized that the process of engaging one's knowledge, namely knowledge of different forms and provenances, never proceeds in a vacuum, but always in specific technological, social, economic and knowledge-related situations, in a climate of certain acknowledged or recognized values, patterns and preferences. Therefore, as a rule, this invariably involves the application and involvement of certain complexes of knowledge, namely the knowledge of not only positive features of technological solutions but also of their negative aspects, risks and potential hazards as well as the engagement of specific patterns, certain value attitudes and preferences, namely values relating to different areas of value structures.

4. A technological solution and, consequently, the outcome of such a solution, i.e. a technological artefact, can be comprehensively summarized by the term *"interface"*, i.e. a system of links and the interface of the inner structure of this solution or artefact and its outer structure, i.e. its external environment, outside circumstances or conditions, including the human and social milieu, authors and users, and – generally speaking – all the other subjects who are somehow affected by the situation resulting from the involvement of a given technological solution. (This particular description, essential for the creation, function and utilization of information technologies, has been formulated by H. Simon). Such a characterization is vital not only for the resultant technological artefact itself, such as its descriptions, properties and functions, but is equally important for the intellectual, information and knowledge-related preparation of the resultant technological solution or technical work. This also concerns all the forms and shapes of assessment during which not only the technological perfection of the given work but also its human-friendliness or rather analogous criteria should be respected.

5. As for the main components of the technological world, its major characteristic lies in what can be called the *"teleological dimension"* (in the sense this particular term was used by N. Wiener) or the *"technological deal"* (a term used by J.-J. Salomon). This primarily involves the fact that goal-directed thinking, reasoning, topics, plans and projects, and also the technological changes made, the technological processes introduced as well as finished technical works have to serve a purpose or give some benefit, that they are, indeed, human constructs primarily focused on human goals, on the task of satisfying human needs and requirements. Furthermore, application of this teleological dimension in the given sense creates not only new goals, claims and requirements to seek and generate new and ever more perfect alternatives in attaining the outlined, accepted and acknowledged objectives, but also spawns new problems and previously unconsidered or completely unknown headaches, difficulties and risks. It can, therefore, be claimed that different elements or units of the technological world have their own destiny or – if we can modify the ancient Latin saying – "habent sua fata instrumenta humana". Nor can we ignore that these fates have their specific rhythm in time, that many of those destinies may be viewed in terms that are usually described as biological metaphors. This means that, metaphorically speaking, we may consider the birth, youth and heydays as well as ailments, ageing and also the demise and death of most types and forms of techni-

cal works and, subsequently, also projects, intentions, implemented processes and fully functional technical works. Similarly, we are justified to talk of a technological time, its directions, rhythms or cycles. In parallel with the direction of time, with the fates of technical works and their developmental cycles we may also follow the rhythms and cycles of technological changes, their nature and level, the sources and motivation of such changes. This is also bound up with discussions concerning the possible typologies of those changes, the role played by technological paradigms, the correctness of analogies with T. Kuhn´s well-known concepts concerning the two types of scientific changes etc. At present, there is a view claiming that such an analogy is rather limited, that we may also consider changes at the different levels of hierarchic structures. When analyzing and considering different technological fates and changes, it is absolutely legitimate – and often very important – to ask to what extent are the different cases of technological changes more or less conditioned by knowledge-related changes and to what extent these changes are also more distinctly reflecting the social, political or religious changes, and also changes in value concepts or value orientations.

6. The technological world is sometimes described as the "second nature". Indeed, this is a nature, even though created by man and human intellect as well as by human values. However, man is an integral part of nature and that is also why the second nature is characterized by the principles and rules of the nature proper. Still, the latter is noted not only for its own paradise on Earth, its beautiful daybreaks, fragrance of its meadows in bloom but also for its wildernesses, inhospitable nooks, its storms and natural disasters. Neither is the second nature only a realm of lasting smiles, it is also a world of mishaps, accidents and major hazards. It is not only a realm of brightness and light but also of shadows and darkness. As early as in the Antiquity, people realized that many new results, discoveries and innovations in the technological world had to be paid for dearly, that they never come on their own and free of charge. After all, this is epitomized by the legendary figure of Prometheus who had to pay a very high price for his courage. (The author of this study has expressed these ideas by referring to the "double face" of technology, or rather the dual face of the ancient Roman god Janus whose one face smiled and the other grimaced, terrifying the onlooker.) The fact that a permanent component of each technological solution and especially of each technological innovation is not only a set of advantages, benefits or otherwise characterized positive values, but also a set of pos-

sible losses, dangers, risks and other negative values, has led thinkers to postulating new orientations in the world of technology whose complex is usually called as the postulate *of technology assessment*. This particular complex covers primarily reasoning in alternatives and weighing the pros and cons of the individual alternatives, laying accents on the "human dimensions" and "human friendliness" of the individual alternatives, complete with evaluation of their feasibility, their economic, technological, environmental, social and health effects, their ethical, cultural and aesthetic values. Therefore, this complex also encompasses what may be termed as an interdisciplinary approach, as an integration of the knowledge of different natural science, technical and humanities disciplines, as an integration of the knowledge – and value – related approaches.

2. Is It Possible to Find an Acceptable Justification of the Philosophy of Technology?

A highly comfortable and basically very cheap explanation may be found to justify the philosophy of technology: There are scholars studying these subjects whose results have succeeded in bringing this discipline a degree of recognition or at least reasonable appreciation. To all intents and purposes, there are fewer of those who are prepared to admit that the results scored in this branch thus far have brought them any major stimuli. Also the kind of argumentation claiming that, after all, there is a philosophy of science, a philosophy of culture and probably also philosophies of other areas of human activities and other analogous forms of reasoning, does not tend to be very convincing either. Actually, it is true that justification of the existence of a phenomenon by merely assuming that such an existence is anticipated by an appropriate type of human behaviour or action may easily call forth many questions and doubts.

When discussing the issue mentioned in the headline of this chapter, we can hardly eschew the question what is, in actual fact, the ultimate goal or meaning of approaching the defined themes we are prepared to perceive as philosophical. Such a query is, doubtlessly, quite meaningful even though answers to it may considerably differ. This also holds true of the analogy to such a question, spelt out in the 1950s by the penetrating thinker and philosopher L. Kolakowski. The question read: "What do the philosophers live off?" While trying to outline at least the main areas of what is the objective or gist of this philosophical approach or the centre of interest of those who regard themselves or are described by others as philosophers, we cannot ignore the following areas in particular:

- the views, ideas or global concepts of other philosophers, namely those of the present and the past,
- the explications, interpretations of such opinions, their systematization or explanation, their arrangement in specific wholes or trends,
- the problems connected with the dissemination of those opinions and ideas, their application to other thematic areas, i.e. with the solution of issues perceived as philosophical "under" certain accepted patterns, schemes or concepts,
- the generation of new opinions, ideas and global concepts that yield or open up new views of the problem areas falling into philosophy's sphere of competence.

The problem areas thus characterized undoubtedly include those associated with learning, with the production of knowledge and knowledge systems, founts of knowledge, with the substantiation, justification or confirmation of knowledge, with language and its means of formulating knowledge etc. Viewed against such a backdrop, this involves traditional epistemological issues which, however, cannot omit the subject of knowledge, attitudes to knowledge, to the utilization or application of knowledge within certain social, cultural and value circumstances and situations. Philosophical issues also cover everything that is or can be the subject of knowledge and, therefore, the topics traditionally termed as metaphysical or ontological. Ontological subjects are closely connected with the issues of the world as a whole, with the sphere involving links of the world and nature, including what we have called the "second nature", man as part of nature and an important subject of the second nature. Seen in this light, this concerns the concepts relating to a global approach to the world and man in it or – as traditionally described in the past – world-view concepts. The issues pertaining to values and value attitudes and hence the problem area concerning value attitudes and value structures also covering ethical, cultural, aesthetic values and other constituents of this sphere, which is sometimes referred to as axiology, cannot be easily separated from global or world-view concepts. At the same time, it is impossible to rule out that some problem areas and the disciplines connected with them have facilitated the origin and relatively autonomous development of some other branches, such as, for instance, ethics, aesthetics, axiology, professional ethics including technological or engineering ethics, the area of technology assessment (TA), the field of social mastering of technology etc.

The problem areas mentioned above, and probably some other spheres falling into the realm of philosophical interests and philosophical approach, in many respects touch upon various entities, phenomena and processes in the world of technology. And when we operate in that world, most of our steps make it imperative for us to have at our disposal knowledge, while quite often we encounter a knowledge deficit, facing the need of acquiring new findings; but this also involves interconnected and interacting types of goal-directed activities and – last but not least – also their results, their actual and possible uses as well as potential impacts of such usage, and – let us add – potential abuses as well. Also these and some other connections have affected many initiatives and procedures applied in numerous already realized steps in the philosophy of technology. It was mainly the philosophy of science, with its accents on the analytical approach, its interest in accentuating the role of language and communication processes that have greatly influenced the hitherto development of major trends in philosophical works devoted to different sectors of the technological world. A special role is assigned in these contexts to another two aspects: the role of values, assessment and a system of acknowledged values of stimulated decision-making and, simultaneously, also the part played by man in different procedures, developmental trends and orientations of leading sectors of the contemporary technological world. This concerns not only man as the creator of a new technology, as a subject utilizing and applying technology, but also man as the being most threatened by the actual and potential trends in the realm of technology. That is why the contemporary philosophy of technology and some new thematic areas lying on the margins of the technical and engineering disciplines, such as technology assessment (TA), the sector of what is called secondary and originally incidental use of some technological innovations (spin-off), must take into account the corpus of important and recognized results scored by those fields of sciences which are characterized as human sciences or the humanities.

For the elaboration of a philosophical approach to the individual areas of the world of technology it is vital to have at one's disposal conceptual sets, traditionally called "conceptual frameworks". Related to the most important areas of the technological world, these conceptual frameworks are formed by the following constituents:

1. the area of technological thinking, which also covers technological knowledge, technological reasoning, technological assessment or evaluation, and the evaluation of the structure associated with the technological world,
2. the sphere of technological actions assuming the form of rational action involving adequate knowledge and values and accompanying the development of a technological solution from its start, conceptual and project preparation, implementation, operation, full-scale utilization, loss of its original functions and inevitable interventions down to its demise or liquidation or its preservation as a major industrial monument,
3. the field of technological artefacts and technical works including short-term interventions, temporary and permanent technological solutions, technological processes and works making up the "second nature" or the already realized or just practicable "possible worlds", their incorporation into broader human, social, environmental and natural contexts that can grow to be a temporary as well as a permanent reality.

3. Technological Thinking

It is necessary to sphere that the sector of technological thinking stands closest to the original meaning of the Greek term "techne" whose role in the etymology of the term "technological" has already been singled out. At the same time, we can hardly overlook the fact that precisely due to this field it is true that the use of the combination of the terms "techne" and "logos" would be absolutely justified in these contexts as well. It is no easy task to characterize the possible specific features of technological thinking, not only because it covers a number of different intellectual processes of a highly diverse nature. We could possibly claim that it operates above the topics of certain specified "possible worlds" (if we can use at all this semantically very inaccurate term, whose closer delineation would call for the assignment of other defining attributes) by means of ideate or conceptual and logical means acquired by man through his intellectual activity over the spheres of the actual world. Furthermore, we can state that technological thinking is motivated by the accessible and available knowledge focused on attaining the outlined goals by known and available means. We can also recall the notion saying that man as the creator, unlike the other living beings, bees for instance, shapes his work first in his head, as an intention, a plan or a blueprint, and only then creates it in reality. This idea was formulated by K. Marx, even though his social and economic conception is fully anchored in the first half of the 19th century, although in some contexts he asserted the view that material production was the source of values.

Technological thinking, especially that aimed at solving more challenging problems, i.e. thinking at the traditional engineering level, basically does not differ from the way of thinking used in those spheres of rational activities that presuppose the involvement of higher levels of knowledge and the rules connected therewith. This applies primarily to the activities of designers, constructors, development and research engineers in R+D centres of large production facilities. In addition to identical aspects of both areas, which facilitate easier transitions of specialists from one area to the other, usually resulting in a certain enrichment in terms of stimulating or searching for new themes or innovation initiatives, some major differences cannot be omitted:

Differences in technological thinking and that employed in other spheres of rational activities lie primarily in their varying goal orientations. In this sense, we may speak of the *teleological dimensions* of technological thinking. These dimensions may then be compared, analyzed, and also evaluated. The objectives of technological activities, which are usually characterized as predominantly practical, include primarily the conceptual, information and organizational preparation and implementation proper of the relevant activity, facilitating solutions of ascertained and recognized problem situations, notably by means of constructs or finds of other human activities, i.e. materials, tools, machinery or otherwise designated artefacts. In a similar vein, research and scientific activity is connected with a specific goal orientation, namely creation, dissemination and extension of knowledge, its justification, confirmation or refutation.

Technological thinking is known to involve different types of knowledge and usually also specific sets of values. But this is also true of other areas of rational thinking, reasoning and deliberation. Even in these contexts one may notice some specific traits of technological thinking. If we understand technological thinking and its outcome as something that may be perceived as an "interface", i.e. a shared boundary between an inner and outer structure of a technical work, i.e. its environs, including the human and social milieu, then the vital types of knowledge also include those concerning the natural, environmental and human conditions. Even simple hand tools were quite justifiably expected to be "human-friendly". And this is all the more true of complex engineering systems in production, power-generation, information etc. Therefore, and this applies primarily to the just mentioned systems, we tend to consider the necessity of assuming an interdisciplinary approach, of integrating different areas of knowledge and, simultaneously, values as well.

Knowledge of adequate procedures of technological solutions involving a well-organized sequence of different activities, in which continuity, choice of suitable means, usable sources or capacities are of great importance, is perceived as a specifically technological type of knowledge. These are the kinds and formulations of knowledge that is collectively described as the "knowledge how" (know-how). These types of knowledge, too, are known to have the nature of generalizations, i.e. they relate to groups of tasks and assignments of a specific target orientation, being formulated as a general instruction for a solution or a procedure of an activity. In these contexts, it is impossible to eschew the questions pertaining to the connections of traditional forms of generalizations, i.e., for instance, scientific laws, confirmed hypotheses or empirical generalizations and general instructions, practical rules some of which may assume the character of algorithms, i.e. general instructions meeting strict formal criteria. It may seem that the knowledge of the type "the knowledge that" can be used to deduce appropriate "knowledge how" in keeping with a general deductive pattern or general rules. In actual fact, it is impossible simply to conceive of a universal scheme and the efficiency of a simple deductive rule should, therefore, be questioned.

Of substantial importance for the world of technology and the kind of thinking, decision-making and reasoning connected with it is *active participation or presence of values* in those areas. This is often closely linked with the teleological dimensions of technological thinking. It is always crucial to attain wanted, intended or defined goals better, more perfectly, easily and quickly. Value orientation also shapes the choice of goal orientations, i.e. we tend to select more favourable goals, goals connected with a higher level of assessment, with more beneficial preferences etc. Since the choice of goals and means results from the decision-making process, seminal importance in this process is reserved for a preferential arrangement of the desirable goals and available resources, capacities and adequate means, recognized and accepted by the decision-making subject. Such a preferential set-up is invariably the product of social, cultural and spiritual developments. It is, therefore, very interesting to compare prevailing preferential arrangements in different stages of history and in selected areas of the technological world. This is demonstrated particularly distinctly by the enormous differences in the preferential arrangements in the development, in the underlying concept of conurbations and in the development of town-planning concepts. Seen in these contexts, a crucial role is played by the spiritual sphere, by notions of the possibilities of afterlife, ideas concerning the importance of necropolises, i.e. "cities of the dead" etc. Analogous results are yielded when analyzing differences in preferences and preferential arrangements that are affected by the development of religious orientations. Historical and social developments also create new goals and new preferential arrangements, for instance ideas pertaining to the role of rest and holidays, of getting to know other values, foreign countries, of staying in beautiful nature, of spending time in the mountains. It is commonly known that the ancient Romans disliked staying in the high mountains and never appreciated such places. As a matter of record, admiration and appreciation of the beauty of high mountains came only much later. These new preferential arrangements tend to give rise not only to new goals but also to the needs of ensuring their technical provision, and, thus, of fostering the development of the so-called tourist industry which is, for some countries, the chief income-earner and a major source of their social, economic and cultural standards.

Of great importance for technological thinking is also what can be called the *inertia of values*. By the term inertia of values we mean respecting the past values even though the actual motives, reasons or conditions of their origin have changed, even though new conditions and new circumstances have arisen. This usually involves values appreciated and respected by the past generations that constituted a kind of natural component of social life. These cover, for instance, respect for certain traditional forms of verbal and written communication in situations where written or verbal form has since been replaced by means of contemporary information technologies. Technical changes and the innovation steps connected with them often result from searching for and creating more perfect, efficient and otherwise better suited alternatives to what we have known thus far, to what has been used in the past and by the past generations, while the former value competence is being somewhat expanded or enriched by the new means of securing this competence.

However, technological changes and major technological innovations often give rise to new value structures or modifications of the existing ones. Recollecting my childhood and holiday times spent with my relatives in a village, I come to realize the enormous value of our Sunday trips in a horse-drawn carriage to a nearby town. The journey included not only a visit to the church but also, later on, to a wayside inn where horses were stabled and which offered a different style of refreshment, encounters with friends and acquaintances and some other situations. It is only natural that a replacement of this type of transport to town by going by car, which also offered the possibility of shopping in department stores and many other possibilities, has actually spawned new values of life. When, after the Soviet-led invasion of Czechoslovakia in 1968 and my forced departure from the Academy of Sciences, I had to find work in a Prague design institute many designers were convinced that the job of a creative designer, standing at a drawing board and a desk and doing demanding calculations, can hardly be replaced by a computer and computer graphics. I did not try to talk them out of their conviction, having no reasonable grounds for such a refutation anyway. Yet, it was possible to save creative designers from dozens of hours of drudgery spent by drawing graphic forms of project documentation simply by feeding all the data in a digital form and by employing a drawing equipment using a digital model of spatial continuum which eventually made the required graphic documentation in a matter of several hours. The same was true of some other tasks, for instance taking different angles when viewing the planned project, making visual simulations of a vehicle's journey through a tunnel that was being designed etc.

It is quite well known how greatly the mass media and modern communications have changed the value structures and preferential arrangement of contemporary citizens, the youngest generation in particular. Many of those changes in preferences seem to be unfolding without attracting greater attention. At the same time, it holds that some of the values, anchored in the social and cultural conditions of the past, are endowed with a certain inertia, still being respected or perceived as permanently valid values.

Comments on the significance and roles played by technological thinking should not omit those constituents of this kind of thinking that open up new possibilities for technological solutions and ideas or images of new technological artefacts. True to say, this does not involve solely repeating the existing, though time-tested solutions, noted for their positive impacts but also those changes which could be described as changes in the technological paradigms. Since each technological solution is associated with a solution of a certain kind of ascertained and acknowledged problem situation, the following types may be distinguished:

- changes in a given situation or rather the emergence of a new problem situation calling for a search for new problem-solving procedures and methods,
- changes in the solution of a given problem situation, finding new procedures and means of the existing solution.

The changes in both above-mentioned categories are or may be more or less radical. It frequently happens that someone finds a new space for the application of the existing procedures, methods and means, and thus also for what is called in the development of science an unexpected set of new possibilities of application (spin-off). At the same time, it is of considerable significance that even this particular component of technological thinking is strongly influenced by the past experience, the hitherto known, time-tested alternative solutions, that images of the past, perceived in a specific light, registered and regarded as important, enter decision-making on current steps as well as measures to be made in the future. For the purpose of giving the technological intelligentsia and especially its youngest generation a better bearing in these issues, the following conclusion should be drawn from the above comments on technological thinking: *Thinking in alternatives,* a willingness critically and responsibly to *assess and evaluate those alternatives* as well as their possible impacts is crucial for each type of technological thinking, which harbours ambitions to ensure its creative character.

4. Technological Action

The actual gamut of different actions we are willing to describe as technological can by no means be separated from the sphere of technological thinking. In most cases, technological actions are launched by certain reasoning and a selected decision-making process or rather a selected alternative procedure. Technological actions are also closely linked with various stages of preparation and implementation of a technological solution or the creation of a technological artefact, with the phases of its use, necessary corrections and modifications, and also with the end of its activities, liquidation or demise of its functions. In a number of key situations this involves a complex of activities whose implementation of goal orientations necessitates the existence of a well-organized sequence of interconnected and succeeding action, while a major role is played by the information links of the individual follow-up activities and, always, by sufficient competence of the subjects of this action as well. Therefore, a network chart is drawn up for many of such complexes, specifying the time and capacity requirements of the individual types of activities and – in more complex situations – demands for required inputs and outputs of such an action, and the actual range and nature of their information links. This is particularly significant for more challenging designing processes, implementation as well as stages of the operation and utilization of large technological projects, such as power-generating complexes, transportation systems, large-scale manufacturing units etc.

The actual sets of technological actions may be classified into several types of action, out of which the following should be regarded as important:

- actions involving the preparation of technological works or projects, notably their design, construction, depiction of their functions or possible impacts,
- implementation and interrelated manufacturing processes, building work, real structures etc.,
- actions associated with the use of technological works or projects, application of their planned or desirable functions, activities connected with technological time, rhythms of their existence as technological works,
- actions involving anticipated as well as unexpected impacts of the relevant technological solutions, risks including unexpected hazards or negative impacts.

It would definitely be possible to enumerate yet other activities that could be termed as technological. The following problem areas are, beyond any doubt, ranked among important and interesting ones for the approaches of an analytically-focused philosopher:

- issues justifying the rationality of different types of technological actions, their teleological dimensions, knowledge prerequisites and their inner action leading to a goal-directed complex,
- analyzing or modelling different activities, their algorithmization (and specification of the limits or obstacles to such automation) for the purpose of automating some partial activities as well as larger units,
- secondary use of time-tested activities for other – or rather newly emerging – needs.

If we are to justify at least in brief the topical nature and the necessity of solving the afore-mentioned problem areas, the following arguments may be presented:

Technological action is rightly characterized as a *rational activity* since it is aimed at attaining specific goals, since it presupposes the application of specific knowledge, experience or rules. It should be added that the quality and value dimensions of the outlined targets should be chosen from the sets of alternatives that are to be evaluated and assessed with a good deal of responsibility. This means that the involvement of appropriate knowledge also includes the application of value criteria bound up with the systems of social, ethical, cultural, health, aesthetic and other values. These value criteria also cover acceptability within the given social, economic and cultural framework, the need of weighing possible developments in a sequence of future, anticipated or potential time intervals. Technological actions often incorporate a multitude of routine actions posing major requirements in terms of time and capacities. That is why an analysis of some of the activities performed by human hands and intellect, their modelling and possible algorithmization to facilitate perspective modelling of some procedures by means of computer technology and computer graphics is of enormous importance.

Rational action linking up to the current and future problems facing the world of technology presupposes that each such action is preceded by an informed, critical and responsible decision-making process, namely on the choice of goals (goals appear in alternatives, too), on the choice of suitable and human-friendly means to be applied, and last but not least also on the rate of acceptability of possible impacts of the preferred solutions or rather methods or ways of coping with the negative aspects of such effects. All the types of technological actions are therefore started off by a certain decision, which covers the choice of acceptable and usable alternatives, namely in the field of target orientations, adequate and available means, sources and capacities. Of substantial and overriding importance for the start of major and usually quite challenging and complex technological projects is what can be called the *"primary decision"* that usually specifies a substantial goal-orientation and the most important procedures and means necessary for its attainment. (Experts in project management tend to speak in such contexts of the start of a design process, i.e. a process called "starting up".)

5. Technological Artefacts

The world of technology is composed of specific man-made units, processes and objects whose creation and utilization involves the application of acquired or attained knowledge and experience. That is why we are entitled to describe the elements, components or means of the technological world as artefacts, i.e. artificially created entities capable of satisfying certain goals, thus also meeting specific expectations. Needless to add, it is neither simple nor easy to answer the question what are the common features shared by simple mechanical units, such as modern cars, washing machines and other similar technical equipment, and finally what has this to do with comprehensive and large-scale technological systems, such as coal-burning or nuclear power stations, transportation systems and other large-scale technological networks. If we want to analyze the creation, properties and functions of various types and kinds of technological artefacts, we have to operate with a number of major terms whose sum total helps us in better capturing the actual nature and significance of technological artefacts. This applies primarily to the following terms that make it possible to depict the character of technological artefacts:

- Technological artefacts are created by man to achieve specific goals, therefore they have their *specific functions* to meet in different spheres of human activities.
- Technological functions are man-created units with a specific *structure*, given by the relations between that whole and its constituents, and also by the interrelations of those elements.
- Structure originates by *concatenation* of those elements, inputs or components, while this concatenation does or may assume a different nature and can be specified by certain rules or limitations.
- It is possible to distinguish an artefact's inner and outer structure, i.e. a structure of its surroundings, environment or conditions among which a defined *interface* operates.
- A technological artefact always has certain *value dimensions*, i.e. it may be accepted as an entity meeting specific functions, it may be evaluated or assessed as something beneficial, helpful etc.

Specific function of a technological artefact describes what can also be termed as its teleological dimension. Man has always produced tools (and that is why he was described as a tool-making animal) since he needed them for a specific purpose, since he could use them for easier achievement of his goals. Due to the same reasons, man is noted for his endeavours to keep improving, increasing the operating power of his tools, gradually also creating ever more complex tools, machinery and technical equipment involving the use of external material and energy resources and then – at a higher level – of automatons. A specific function of different kinds and types of tools and technical equipment similar to tools has also come to be used for the designation of the latest of such tools and equipment. While the oldest types of tools have in most languages their own original names, to denote the newly developed technical equipment we usually use designations expressing their specific function. So, while still using a knife and fork in the kitchen, we also have a dishwasher, an automatic washing machine and many other household appliances.

The term "*structure*" is of great importance for understanding the specific functions of technological artefacts. Under the term structure we understand a composition of a larger unit from its partial elements, constituents or parts, which gives rise to new properties or functions that its individual elements, constructional parts or partial components do not and cannot have. Some thinkers speak in this context of emerging new specific functions (J. Piaget). The process of creating structure from elements which then make up parts of larger units poses a number of problems that are an accompanying feature of preparatory activities, of the development or design process. This relates primarily to the issues indicated in the questions given below:

- Is it possible to extend the structure of a given type, to improve or expand the existing structure?
- Is it necessary to strive for a design, initial study or project of a new structure?
- Is it possible, permissible or acceptable to seek the substitution of some new elements with the aim of improving or prolonging the function of some structures? (This question is of particular importance for combining the elements of different nature and provenance.)

What are the main ways, methods or procedures conducive to reducing risks or other threats in the activities of some structures? (This issue constitutes an important and crucial problem in ensuring and enhancing the overall reliability of technological artifatcs.)

The term "*concatenation*" was originally used in linguistics and then in logic and mathematics to denote a permissible combination of two or more signs which gave rise to a new category of signs with a new meaning different from that of the original isolated signs. There are different examples of concatenation in the nexus between subject and predicate in logic, a noun and verb in linguistics, giving rise to a sentence with a new meaning, in the link between operators and suitable operands etc.

Analogous connections between initial elements or components whose concatenation generates a new technological structure may also be considered in the field of technological activities involving the creation of a new structure, models of this structure and their implementation. Even in these spheres we are entitled to envisage the presence of certain rules bound up with some specific types of concatenation. A key position in the systems of such rules is occupied by the principles of permissible concatenability. In technological terminology, we tend to speak in similar contexts of compatibility, of the limits to concatenability, to compatibility etc. Nor should we ignore the fact that many rules for concatenation, preferred concatenations, for admissibility, acceptability or other analogous characteristics are more or less intuitively envisaged rather than explicitly formulated. We can, therefore, emphasize that in the technological world too there are important rules having analogous properties and functions just as the syntactic rules in the sphere of communication and in the applications of natural and formalized or graphic languages. Also, we should not overlook that in addition to inertia of certain value structures, there exists – in the development of the world of technology – an inertia of rules according to which certain types of structures are formed and perpetrated. There were specific rules of written correspondence, starting with an address of the recipient and ending with final greetings. Most communications sent by electronic mail more or less respect very similar rules. Likewise, the technological artefact, called the robot in keeping with Karel Èapek´s coinage, usually has two eyes, two arms and other parts corresponding to the structure of the human body.

A major feature characterizing technological artefacts is the *interface*. In practical terms, this means that an axe is a suitable tool not only because its metal wedge is connected to a wooden handle but also because there are people who know how to take up an axe and chop previously cut timber that was prepared for processing into fuel. Similarly, a car is a reliable means of transport not only because its body is equipped with a combustion engine, a suitable steering and transmission systems and other necessary components of a roadworthy vehicle, but also because its environment is adequately prepared for its function. This covers the construction of a network of roads, filling stations and many other facilities, skilled drivers included. Commenting on the importance and role of the interface, we can hardly omit mentioning the naive notion that "technology is a panacea", that new technology will immediately solve all the problems and difficulties, and that by introducing the Internet in our school system people in those facilities will be suddenly enlightened by necessary knowledge. The topic of the interface and its seminal significance in all the innovation processes may also be aptly illustrated by the errors and defects committed in the recent past, when. e.g. housing estates had been completed without finishing the necessary engineering and civic amenities, a reliable transportation system etc. A case in point documenting how utterly ignored was the impact of the interface in the past was the construction of large manufacturing complexes in the era of totalitarianism and the system of monopoly power in this country in localities where raw materials had to be transported over great distances and where power-generating facilities were lacking. Seen in this light, an interface constitutes a wide-ranging network of links of a highly varying nature, i.e. material, energy, information links, sufficient user abilities and skills, and qualification prerequisites, regard for ecological concerns and for the environment, for natural conditions etc. The need to devote great attention to the area of relations and links should, therefore, be projected into primary decisions on technological changes and innovations; they have to become an integral part of their conceptual and project preparation. Equally significant are also all the forms of communication processes instrumental in safeguarding their qualified use and control. This should apply not only to what are called routine situations but also to potential emergency or risk situations. Communication processes incorporate what is termed the social mastering of technology. Due to these reasons, the author of this study has given an affirmative answer to the question whether we are justified to speak of the semantics of technological artifacts.

Claiming that a technological artefact always has a *value dimension* is connected with the actual nature of decision-making processes that accompany the entire "lifetime of a technological artefact", starting with reflections and decision-making on its necessity, expediency, costs of its creation, the uses or benefits its application may yield that will eventually be reflected in the primary decision-making and in the subsequent design process. It is an absolute simplification to claim that the procedures of a comprehensive, multicriterial assessment of a proposed technological solution and its potential impacts may be somehow incorporated, in a one-off fashion, into the process of preparatory work. Assessment and evaluation constitute a major component of these decision-making processes, which actually accompany the technological work or project from its genesis to its demise or liquidation. Some of the contemporary discussions confirm that the most difficult problems appear in connection with the liquidation of technological works, with efforts to eliminate or at least lessen their adverse effects. This also applies to the current difficulties and the related debate on the disposal of nuclear waste from nuclear power plants, the liquidation of the negative health effects of intense fertilization and chemization of present-day food production, the mass-scale application of antibiotics in what is called livestock and meat production etc. Until now bitter clashes have been fought between economic interests on the one hand and health and ecological criteria associated with environmental protection on the other in tackling the problems posed by the greenhouse effect, the ozone layer and other analogous issues involving some of the irreversible processes caused by the impact of the current technological world.

Literature

Bunge, M. 1974. *Technology as Applied Science.* In: F. Rapp, ed., Contributions to the Philosophy of Technology, D. Reidel Publ. Comp., Dordrecht, Boston.

Ellul, J. 1963. *The Technological Society.* Vintage Books. New York. Translated from the French.

Heidegger, M. 1962. *Die Technik und die Kehre.* Verlag G. Neske, Pfullingen

Kornwachs, K. Hrsg. 2004. *Technik, Systém, Verantwortung.* LIT Verlag, Münster

Kotarbiñski, T. 1874. *Instrumentalization of Actions.* In: F.Rapp, ed., Contributions to the Philosophy of Technology, D. Reidel Publ. Comp., Dordrecht, Boston.

Lebeau, A. 2005. *L´engranage de la technique.* Gallimard, Paris.

Lenk, H., Maring M. eds. 2001. *Advances and Problems in the Philosophy of Technology,* LIT Verlag, Münster.

Lenk, H., Moser S. 1973. *Technik, Techne, Technologie.* Verlag Dokumentation, Pullach.

Piaget, J. 1972. *The Concept of Structure.* In: Bunge M., Bar-Hillel Y., Tondl, L. et al., Scientific Thought, UNESCO, Paris.

Rapp, F. ed. 1974. *Contributions to the Philosophy of Technology*, D. Reidel, Dordrecht, Boston.

Rosenblueth, A., Wiener, A., Bigelow, J. 1943. *Behavior, Purpose and Teleology.* Philosophy of Science, vol. 10, pp. 18–24.

Salomon, J.-J. 1992. *Le destin technologique.* Balland, Paris. Czech edition. *Technologický údìl*, Filosofia, Praha, 1997.

Simon, H.A. 1969. *The Sciences of the Artificial.* The MIT Press, Cambridge, Mass.

Tondl, L. 1968. *Der Januskopf der Technik.* In: Akten d. XIV Internat. Kongresses der Philosophie. Bd. II. pp. 570–577.

Tondl, L. 1998. *Technologické myšlení a usuzování. (Technological Thinking and Reasoning).* Filosofia, Praha. (In Czech. German edition: *Technologisches Denken und Schlussfolgern.* Edition Sigma, Berlin 2003).

Wiener, N. 1954. The Human Use of Human Beings. Czech translation: *Kybernetika a spoleènost.* Nakl. ÈSAV, Praha 1963.

The Praxiological Concept of Technology[1]

Wojciech W. Gasparski
Leon Kozminski Academy of Entrepreneurship and Management
Warsaw, Poland

1. Introduction

The present paper attempts to define technology and typologically distinguish its branches, including the technical sciences, as the subject of technical sciences studies. Who knows, it may actually be more appropriate to have these studies encompass not just the technical sciences, but technology as a whole?

The starting point for these reflections will be the action-based approach to technology, an approach that has often been put forward by Zbigniew Wasiutyñski [6, 7, 8] and Bohdan Walentynowicz [5] as well as the author of the present paper [9]. Additional encouragement for such an approach to technology is provided by Tadeusz Kotarbiñski:

> In the huge field of investigation that has science as the object of its study, one can distinguish two main branches, namely studies of science as a *set of theorems* and studies of science as a *set of actions*. Science, after all, is a part of culture, therefore the same kind of duality appears in relation to science as emerges in relation to other component spheres of culture, e.g. technology, economics, law, literature, art; in all these, there is a need to distinguish between a *set of actions* and a *set of achievements*. This difference becomes prominently visible in *technology*, where we have artificial substances, machinery and structures on one hand, and on the other – conceptual and execution work producing those same objects [3].

Concluding these introductory remarks, it is worth mentioning that the literature includes definitions of technology that are a consequence of the object approach, thus reducing the meaning of this concept only to machines, equipment etc. This paper does not set out to polemicize with such definitions. The goal here is to present terminological proposals that appear to be more convenient for the study of science or the study of technology, and primarily for the emerging discipline called the *methodology of technology*, in particular with regard to problems related to design.

2. Human actions and their instrumentalization

Let us denote the set of all human *actions*[2] (Box 1) as $\{A\}$. This set has the qualities of a universal set. From this universal set, let us separate the set of *preparatory actions* (Box 2) $\{P\}$. From the set of preparatory actions, let us separate the subset $\{I\}$ whose elements, or the actions making up this subset, are defined as follows: the overriding goal of every action belonging to $\{i\}$ is *instrumentalization* (Box 3) of the actions forming the universal set $\{A\}$, Fig. 1.

> An *action A* is an intentional, conscious human behavior compatible with the will of the acting subject.

Box 1

> Action A_1 is a *preparatory action* with respect to action A_2 always and only when A_1 precedes A_2 in time and when action A_1 enables or facilitates the execution of action A_2.

Box 2

> *Instrumentalization* means economizing effort through bringing about mediated action by introducing a tool – *technical object* – between the worked object and the acting subject.

Box 3

This does not mean to say that every action belonging to $\{I\}$ has as its overriding goal fulfillment of the necessary condition of instrumentalizing all the actions comprising $\{A\}$. It is sufficient to fulfill the necessary condition of instrumentalizing a single action or a group of actions making up $\{A\}$.

Figure 1

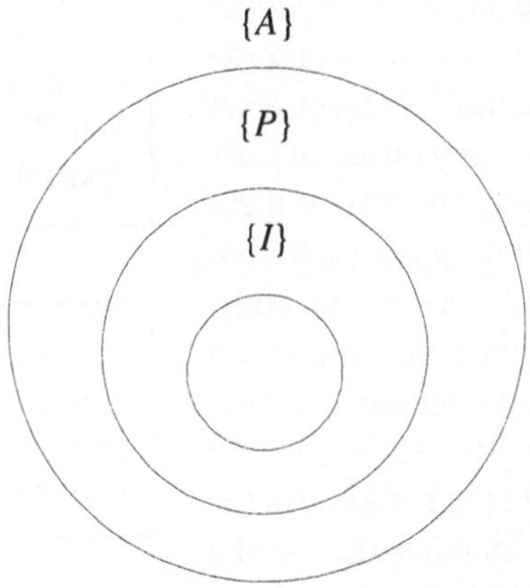

Let us group the actions included in the subset {I} according to similarities as described below, and arrange the groups in the order of succession in time. Let us label the groups – types of actions, from right to left, as {T}, {T-1}, {T-2}, ..., {T-k}. In Fig. 2 illustrating the above sequence, on the right-hand side let us add an element symbolizing the universal set {A}, and on the left-hand side the element {S}, which will be explained later.

At the end of the described sequence will be actions whose *external goal* (Box 4) is identical with the overriding goal of the actions included in {I} as presented above. These actions, labeled as {T}, will be called *technical actions*.

Figure 2

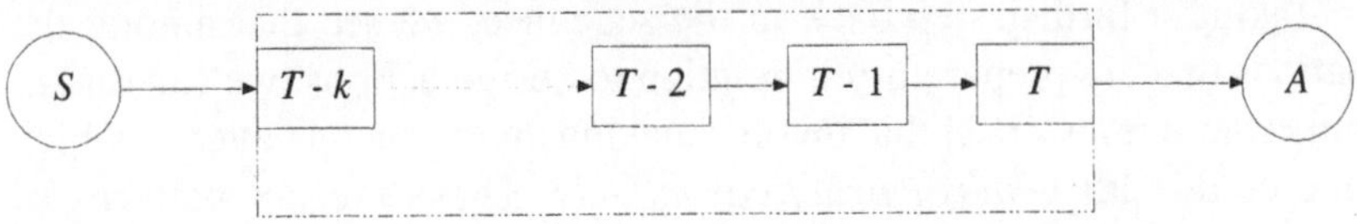

The *internal goal* (Box 5) of technical actions {T} is to make instruments (apparatus). There are two kinds of such making:

- first, shaping various kinds of materials so as to result in an object called a *technical object* [Wasiutyński 8];
- second, ensuring this technical object's *functioning*.

Let us call the former type of making *production*, and the latter – *technical operation*. Let us explain this on an example. Producing a technical object, a car for instance, requires many different parts to be made from different materials, these parts to be combined into certain self-contained entities called units (engine, body, seats etc.), and finally the units to be joined together into one whole – the car. For a car to be able to function, you also need to fill its tank with fuel, load the battery, fill the tires with air, provide maintenance, make repairs etc. – and this is what we have called *technical operation*.

External goal is a further goal realization of that exceeds the system within which the action is performed.

Box 4

Internal goal is an adjacent goal realization of that is performed within the system of a given action.

Box 5

Technical actions of the kind described above, comprising production and technical operation, is "technology in the narrow sense". In other words, *technology in the narrow sense* comprises actions whose internal goal is the production and technical operation of technical objects, while their external goal is fulfillment of the necessary condition of instrumentalizing the actions forming the universal set $\{A\}$.

Now, let us take a step back in the previously built sequence. Prior to the actions counted among technical actions $\{T\}$, there are the earlier actions $\{T\text{-}1\}$. These actions are preparatory actions in relation to the actions $\{T\}$, and pre-preparatory in relation to the actions $\{A\}$. These are actions that involve making *models* (patterns) of technical objects and actions. Let us call these actions *design actions*, and the models *detailed practical models*.

Taking a further step back in the sequence, we see that among the actions that are preparatory in relation to design actions, we can distinguish the actions $\{T\text{-}2\}$ that involve making more general patterns, which one could call *general practical models*. These are not patterns of a single technical object or an action, but patterns of a whole *family* of technical objects and actions, Fig. 3.

Moving back step by step in the sequence of actions, we can distinguish actions that involve making increasingly general models, ones that from some step can no longer be called practical models, but are *theoretical models*. In relation to the actions $\{A\}$, each of these actions is a pre-preparatory action of an increasingly higher order i ($i = 1, 2, \ldots, k$).

Figure 3

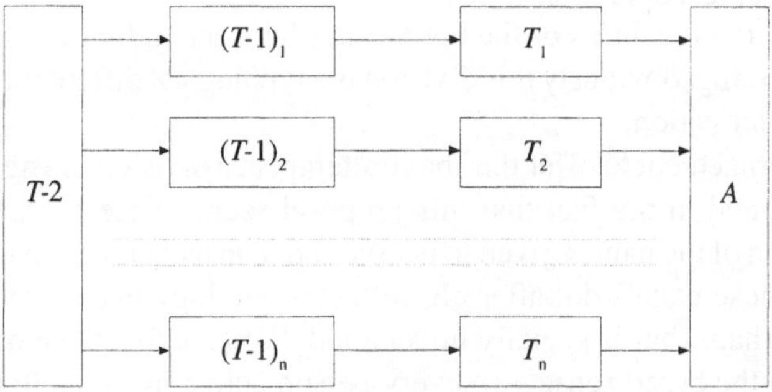

All the actions outlined above that are pre-preparatory in relation to the actions $\{A\}$, together with the technical actions $\{T\}$ comprise *technology in the broad sense* or, as proposed here, *engineering*.

Figure 4

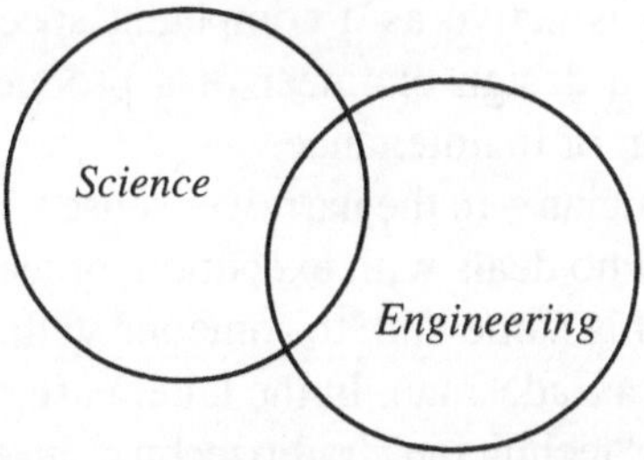

With regard to engineering in this sense, that part of it which encompasses preparatory actions consisting in making general models comprises the *technical sciences*; this part can be called *cognitive engineering*. That part of engineering that is contained between the technical sciences and technology in the narrow sense can be called *conceptual engineering*. This name also encompasses the element of technical design that is linked to developing practical models, the element known as engineering art.

Thus, what we have is engineering as the general name for *engineering actions*, encompassing

- technical sciences, or cognitive engineering
- conceptual engineering, or technical design
- technology in the narrow sense, or technical (execution) actions, which in turn include:

- production
- technical operation

Due to the fuzziness of the borderlines between individual segments of engineering, obviously this division is a typological differentiation and not a classification.

Encouragement to offer the above attempt at a division of engineering is also found in the fact that this proposal seems close to the general perception of the names given to people active in technology in the broad sense. These names do, after all, reflect a typology that is not written down perhaps, but is socially understood. We give the name of *technician* – in the broad sense – to every person active in technology in the broad sense, regardless of whether they are a worker or a technical university professor. We give the name of *engineer* to those from among the above-mentioned technicians who have a higher education in and knowledge of each of the three segments of engineering: technical sciences, design and execution. Depending on the proportion in which an engineer is familiar with those segments of engineering, and in which segment that person is active as a competent specialist, we speak of a research engineer, a design engineer or a production engineer who supervises production or maintenance.

The name of technician – in the narrower sense – is given to an expert in technical actions who deals with execution, or technology in the narrow sense. We give this name both to someone with a technical secondary education and to a tradesman. In the latter case, we add an explanatory term to the word "technician": radio technician, electrical technician etc. or, when the name of the occupation does not feature the word "technician", e.g. mechanic, turner, fitter etc., we use a more general term – technical worker.

Returning to the object of interest for technical sciences studies, having called these sciences cognitive engineering and having delineated one of the borders, namely the border between them and conceptual engineering, let us now try to map out another border. This time, it is the border between technical sciences and natural sciences.[3]

It would seem that this border can be marked in the place where the next step back in the previously described sequence of actions will cause us to find ourselves outside the sequence of actions whose overriding goal is fulfillment of the necessary condition of instrumentalizing the actions $\{A\}$ (segment on the far left marked as S in Fig. 2). This is the sphere of actions whose overriding goal is simply *cognition*[4] and about which we cannot say that they comprise preparation for instrumentalizing the actions $\{A\}$, though indirectly they provide a cognitive foundation for those actions.

3. The Concept of Action Instrumentalization and Its Quality and Efficiency

Let us consider a *technical object* as a set of sets. We will call these sets *design parameters*. The elements of these sets are the design parameter values. Let technical object α be a set of size k, which means this object, is described by k design parameters. Working on a design of a specific technical object, the designer selects certain sets of values from the individual sets (design parameters). This involves calculation, determination on the basis of tables, graphs, standards etc., of the nominal values of the design parameters, and then selecting the environment – called tolerance – for each of the nominal values.

For each ith design parameter, we have a specified nominal value x_i and environment X_i defined by the values: lower x_{il}, upper x_{iu}. Designing results in detailed practical model $M_\alpha = \{X_1, X_2, ..., X_k\}$, on the basis of which individual specimens of technical object α (articles/products) are made, being an ordered set of design parameter realizations $W_\alpha = (w_1, w_2, ..., w_k)$, where w_i denotes the realization of parameter X_i.

The *quality* of a technical object is defined by *accuracy* and *rationality*. The accuracy D of the execution of technical object a is defined as follows:

- $D = 1$, if $w_i \in X_i$ for all i
- $D = 0$, if $w_i \notin X_i$ for even one i

Making technical object α (production activity p) and using it (technical activity v) requires outlays $N_{\pi\alpha}$ and $N_{\upsilon\alpha}$. Dividing the utility effect ΔZ_α, expressing economy of resources in an action achieved thanks to instrumentalization of that action, by the sum of outlays on the technical object $(N_{\pi\alpha} + N_{\upsilon\alpha})$, we obtain a quotient that is the *instrumentalization efficiency* E_α of technical object α:

$$E_\alpha = \Delta Z_\alpha / N_{\pi\alpha} + N_{\upsilon\alpha}.$$

Assuming that at least a correlation relationships exist between the variables of the instrumentalization efficiency E_α of technical object a and the design parameters X_i of that object, then – on the basis of the known values of the design parameters x_i for different variants of the model of the technical object – we can determine: $E_1, E_2, \ldots, E_q$.

The condition for recognizing a variant of a technical object to be efficient in the praxiological sense is for $E_\alpha > 1$.

The rationality R of models of a technical object is defined as follows: $R = 1$ for $E > 1$; $R = 0$ for $E < 1$.

Quality Q is the Boolean product of rationality R and accuracy D:

$$J = R \wedge D.$$

Quality as defined above assumes the following values:

$\wedge$	$D = 0$	$D = 1$
$R = 0$	$Q = 0$	$Q = 0$
$R = 1$	$Q = 0$	$Q = 1$

The measure of the preciseness of organization of a complex object (the degree in which that object is a system – its *systemic nature*) is the *organizational effect*. This could concern:

- a team of people – then it is an *organization*
- the pair "human – technical object" – then it is *instrumentalization*

This is explained by the fact that the pair "human – technical object" as a complex object is a kind of organization, and its measure of organization (*systemic nature*) is the equivalent of the organizational effect called the *instrumentalization effect*, and instrumentalization – as mentioned earlier – means using a technical object in an action performed by a human.

The (organizational) *effect of instrumentalization* ε_i is the difference between resources used in non-instrumentalized action Z_0 and instrumentalized action Z_1:

$$\varepsilon_i = Z_0 - Z_1$$

If the instrumentalization effect achieved as a result of using technical object α in an action is $(\varepsilon_i)_\alpha$, while the instrumentalization effect achieved as a result of using object β in that action is $(\varepsilon_i)_\beta$, then the instrumentalization effect growth

$$\Delta\varepsilon_i = (\varepsilon_i)_\alpha - (\varepsilon_i)_\beta = Z_1 - Z_2 = \Delta Z$$

Instrumentalization effect growth as defined above is the *utility effect* [7] of a technical object. The utility effect expresses economy of resources in an action, achieved thanks to instrumentalization of that action with a given technical object.

4. Conclusion

In conclusion, one hopes that even with any controversy it involves, this attempt at introducing some terminological order, thanks to praxiological concepts, shows there are no grounds for opposing science to technology and vice versa. Such a tendency seems to have resulted from the lack of an accurate action-based definition of technology, and from the simultaneous diversity of meanings assigned to technology, and thus substantial arbitrariness in using the term.

In the approach offered in the present paper, engineering and science are fields of action that partially overlap, Fig. 4, where the common part is the technical sciences, i.e. cognitive engineering.

It is clear from the presented method of building a definition of technology (in the broad sense) and its branches that one of the features of actions counted among engineering actions is their *being directed* towards the practical goal of instrumentalization of actions. This suggests that the popularity enjoyed by the term "technology" outside engineering is caused – among other things – by the understanding of technology as this kind of *directing* towards practice separately from the other features distinguishing technology as a field of engineering actions.

Notes

1. This paper is a revised version of a paper published in *Zagadnienia Naukoznawstwa* (Science Studies Issues; in Polish), No 17:117–122.
2. The terms: "action" and "preparatory action", "instrumentalization" etc. are used as they are understood in praxiology, see boxes.
3. Or rather: mainly the natural sciences, as cognitive engineering goes beyond them, to mention but mathematics.
4. Neither about which Hans Selye wrote that neither the glory science could provide nor its potential usefulness are its only assets, and certainly not its usefulness in the sense that most people understand it. The main benefit from fundamental research, the same as from a rose bloom, a song or a beautiful landscape, is that it gives us pleasure. Every scientific discovery reveals a new harmony of the laws governing Nature for our passive enjoyment. [Hans Selye, 1964, *From Dream to Discovery*, McGraw–Hill, New York].

References

Kotarbiñski, T. 1965. *Praxiology: An Introduction to the Science of Efficient Action*, Pergamon Press, Oxford.

Kotarbiñski, T. 2003. Mutual Links Between Different Aspects of Studies of Scientific Output, in: T. Kotarbiñski, *Collected Works: Praxiology Part 2* (in Polish), Ossolineum, Wroc³aw, pp. 561–571.

Kotarbiñski, T. 2003. On the Usefulness of Praxiology, in: T. Kotarbiñski, *Collected Works: Praxiology Part 2* (in Polish), Ossolineum, Wroc³aw, pp. 186–189.

Pszczo³owski, T. 1974. *Concise Encyclopedia of Praxiology and Theory of Organization* (in Polish), Ossolineum, Wroc³aw.

Walentynowicz, B. and Wasiutyñski, Z. 1962. Praxiology and Technology (in Polish), *Materia³y Prakseologiczne*, No. 4, pp. 207–228.

Wasiutyñski, Z. 1962. *Technology, Its Branches and Method* (in Polish), PWN, Warsaw.

Wasiutyñski, Z. 1962. *Analyzing Utility Effects in Technology* (in Polish), PWN, Warsaw.

Wasiutyñski, Z. 1964. *Analyzing Utility Effects in Bridge Engineering* (in Polish), PWN, Warsaw.

Wasiutyñski, Z. and Gasparski, W. 1969. From Research Results to Practical Application in Technology (in Polish), paper presented at a conference of the Polish Academy of Sciences' Laboratory for General Problems of Work Organization, Warsaw 1966, in *Prakseologia*, No. 28, pp. 185–209.

Wójcik, T. 1963. The Subject, Goal and Task of Praxiology (in Polish), *Problemy Organizacji*, No. 1, pp. 9–57.

Zieleniewski, J. 1966. *Organizing Human Teams* (in Polish), PWN, Warsaw.

Positive Feedback and Praxiology: Path Dependence in Action

W. E. Herfel
University of Western Sydney
Australia

"… the growth of culture creates special compulsory situations. Certain elements of culture become a burden which must be got rid of. One of the most common phenomena here is the fact that a given element of culture has its inertia, i.e. occupies a place, and an effort is needed to replace it. … An acquired habit persists, even when a new habit is already needed. Progressive achievements become, through their inertia, stumbling-blocks to progress. And here is a new paradox in the dynamics of progress – the paradox of the apparent conservatism of innovators with respect the later innovators who have availed themselves of the experience of the earlier and have introduced better improvements in their own field, not burdened by earlier conceptions. The weight of culture increases with the accumulation of its elements. More and more things have to be learned and remembered if an individual wants to be abreast of the latest developments, and *a fortiori* if he wants to triumph in competition and, in general, to advance."

Tadeusz Kotarbiñski (1966, p. 197)

1. Introduction

In the middle of the last century Tadeusz Kotarbiñski identified a crucial aspect of technological progress. Recent developments in the modeling of economics of technological markets allow us to clarify and extend Kotarbiñski's insight. This paper will outline the results of these models and apply their results to analyse this "burden of the inertia of culture."

According to economic orthodoxy, technology markets, unencumbered by external interference, will deliver ever-improving technological solutions to life's difficulties. Entrepreneurial engineers will naturally use their ingenuity to improve technology. Well-informed consumers will naturally purchase the best available products. In this way free markets facilitate an efficient incremental search through technological possibility space yielding in the long run the best of all possible technological devices. Like many other critics, economist Brian Arthur was disturbed by a lack of fit between the neo-classical theories and the available economic data. In particular, he was troubled by several issues: Why do technology markets so often come to be dominated by one or two producers? Why do the goods they produce fall so short of optimal technologies? Why do outdated technologies persist in the face of superior alternatives? These questions seemed to apply especially in emerging high technology (hi-tech) industries that have come to dominate modern economies. Disappointed with orthodox economic analysis with its fundamental assumption of negative feedback in the guise of the "law" of diminishing returns, Arthur was convinced that positive feedback played a far greater role in markets than orthodox economic theory leads us to believe. He thus provided answers to this sort of inquiry by embarking on the study of positive feedback models, applying them to technology markets.[1] We will explore one such model next.

2. Polya Processes

A simple statistical model of positive feedback was developed by mathematician Georg Polya. Imagine that we have an infinite supply of balls of two colours, say red and blue. Take one of each and place it into an urn of infinite capacity. Now withdraw a ball at random from the urn. Replace the chosen ball, along with another ball of the same colour to the urn. Repeat the process indefinitely. What is the result of this process in the long term? Since the proportion of blue balls equal one minus the proportion of red balls, we only need to discuss the proportion of red balls. Polya proved that the proportion of red balls converges on a particular number, x, as the number of iterations of the process approaches infinity. But what is x? Polya further proved that x is a random variable ranging uniformly from null to unity! (Arthur 1994, p.36)

To see what this means imagine carrying out this process repeatedly. In the long-term a constant proportion of red and blue balls will occur.

Sometimes the urn will approach a state of containing all red; sometimes no red, but more often it will be some proportion in between. Each proportion is of equal probability. Arthur's (1994, p. 37) computer simulations of Polya processes reveals the dynamical details. The process exhibits positive feedback; picking red initially increases the odds of picking red subsequently. However, early on the number of balls is small hence the chances of picking a non-majority colour is relatively high. Hence, during the first iterations of the process the proportion of red balls wanders. However, it is what happens in these early iterations that determines where the long-term convergence will be. Eventually there are enough balls in the urn that the odds are high that subsequent selections will reflect the present proportion. This tendency increases with the number of iterations. After a crucial number of iterations the pattern is set; little fluctuation in proportion is observed.

Arthur points out two important features of this positive feedback: path dependency and lock-in. There is no way to predict *a priori* the result of any given instantiation of the process. In fact, Polya proved that each proportion is equally probable. Path-dependence is the property that it is the early iterations of the process, rather than some end toward which the process is directed, that determine the long-range outcome. However, the long-term dynamics is locked in; once the proportion of red balls is established it will remain constant.

This model (albeit grossly simple) accurately captures important aspects of positive-feedback markets, e.g. markets whereby consumer choice is influenced by current market share.[2] Arthur identifies the intensity of the knowledge basis required for high technology industries as the chief source of positive feedback in their markets. High up-front investment is required for research and development to produce hi-tech items, whilst production cost are relatively low. "Moreover, experience gained with on product or technology can make it easier to produce new products incorporating similar or related technology" (Arthur, 1994, p. 3). Perhaps the strongest source of positive feedback in knowledge-based products is that they require compatibility: "when one brand gains a significant market share, people have a strong incentive to buy more of the same product" (Arthur, 1994, p. 4). As Arthur points out individuals need to be able to exchange information with others. But as well each individual user has a huge investment in knowledge of how to operate the technology and needs later instantiations of a given technology to be compatible with earlier models. How many of my readers have regularly switched between IBM and Apple personal computer platforms over the past two decades? Arthur's simple model of positive feedback can explain why the 12-hour, clockwise clock face came to dominate European clocks, why VHS swamped Beta in the VCR market, and why for several years the Apple Macintosh held a fairly constant share of the PC market. Arthur (1994, chs. 3 and 10) along with statistician colleagues, Yuri Ermoliev and Yuri Kaniovski extended Polya's result to a wide class of such processes.

3. QWERTY

Perhaps the most spectacular example of technological lock-in is the QWERTY keyboard. Anyone who has ever used a computer is familiar with this piece of technology. And I do not think that many have failed to speculate on why the letters are arranged as they are. With a little reflection on the practice of touch-typing and only rudimentary knowledge of the frequencies of the occurrence of letters in English[3] prose, many have guessed that the origins of the QWERTY keyboard are based on *slowing the typist down* to make early typewriters mechanically viable. However, this guess cannot be correct: when QWERTY was invented fast-touch-typing had not yet been conceived! The true story of QWERTY is instructive.

Economic historian, Paul David, has detailed the development of QWERTY in several contexts. I will not repeat the whole account here but will touch on some highlights. A precursor to the QWERTY arrangement was devised by C.L. Sholes, a mechanically inclined printer who was the fifty-second person to patent a "typewriter," in the 1870's. His design was not particularly successful owing mainly to the fact that it was an up-stroke machine. In this design the keys struck the paper from below the paper carriage which held the paper flat. The printing process was invisible to the operator making the tendency for keys to jam particularly annoying. A jam could go unnoticed. When the operator lifted the carriage to examine the fruits of his labour only then would a string of repetitions of the stuck letter be discovered instead of the intended page of type. After six years of trial and error tinkering to overcome this problem, the initial alphabetical arrangement had evolved into an arrangement close to QWERTY (David, 1986, pp. 35–36).

Sholes failed to bring a workable machine to market, but the arms manufacturer Remington & Sons took over manufacturing rights to the

"Type Writer", and eventually sold as many as 5,000 of the Remington Type Writer Model Two by the beginning of the 1880's. "QWERTY thus had evolved primarily as the chance solution to an engineering design problem in the construction of a typewriter which would work reliably at a rate significantly faster than a copyist could write" (David, 1986, p. 36). David goes on to detail subsequent explosion in technological development of typewriter technology. Many competing designs emerged and several innovations resulted including design with a visible printing point. Amongst the design innovations were efficient keyboard layouts including one placing "DHIATENSOR in the home row, these being ten letters with which one may compose 70 percent of the words in the English language" (David, 1986, p. 38). David (1986, p. 38–-39) continues,

> The beginning of the typewriter boom in the 1880's had thus witnessed a rapid proliferation of competitive designs, manufacturing companies and keyboard arrangements rivalling (sic) the Sholes–Remington QWERTY. Yet, by the middle of the next decade, just when it had become evident that any micro-technological rationale for QWERTY's dominance was being removed by the progress of typewriter engineering, the US industry was rapidly moving toward the standard of an upright front-stroke machine with a four-row QWERTY keyboard that was referred to as 'The Universal'.

The rest of the story details how QWERTY did indeed become universal. The story is intriguing; it includes classes delivered at the YWCA in New York City, debates in journal *Cosmopolitan Shorthander*, and a contest won by "all finger" typist Frank McGurrin. Whilst the details of this case are not important to recount here (although I refer the reader to David's wonderful account) the lesson is crucial: QWERTY is a classic case of path dependent lock in of technology. And David's argument is that such cases can only be explained through recourse to the historical contingencies leading to the technology in question.

As Arthur has argued when positive feedback dominates a market the resultant structure has little to do with the technological efficiency of the competing products. The results of high technology markets are path dependent: a series of seemingly small events lead to particular devices becoming locked in to our everyday lives. David identifies three factors that caused QWERTY to become locked in: technical interrelatedness, economies of scale and quasi-irreversibility of investment. (David, 1986, p. 41) These are in Arthur's terms sources of positive feedback. In general, one is tempted to state that it is positive feedback that explains why

QWERTY came to be *the* typewriter keyboard. And it explains why this dubious piece of technology has persisted into the computer age. However, this way of putting it is misleading. Positive feedback explains how inefficient devices persist whilst efficient ones end up as museum pieces, but if we want to understand QWERTY our only recourse is to history.

Some have cast doubt on the validity of the path dependence thesis. Stan Liebowitz and Stephen Margolis have written numerous articles aimed at discrediting what they interpret as a fundamental distrust of markets peddled by advocates of the positive feedback models of markets. David (1986, p. 33) cites August Dvorak's studies conducted in the 1940's where he claims that efficiencies gained by the adoption of his keyboard would "amortize the cost of retraining a group of typists within the first ten days of their subsequent full-time employment." Dvorak's layout was offered as an option on the Apple IIC at the flick of a switch with Apple claiming a 20–40% increase in typing speed (David, 1986, pp. 33–34). Liebowitz and Margolis (1996) convincingly argue that Dvorak's studies were hopelessly flawed, hence perhaps the markets have not been so hopelessly inefficient in delivering us QWERTY after all. The evidence that *they* cite in favour of QWERTY refers to typing contests *taking place in the late 19th century.* I would suggest the jury is still out on the extent of the inefficiency of QWERTY. And perhaps we can grant that QWERTY is indeed good enough.

Nevertheless, we are still left with the question of why QWERTY in the first place. As Arthur (1994, p. 15) points out, assuming 40 or so keys, there are 40! or 10^{48} possible ways to lay out a typewriter keyboard. The market decided on one. Certainly most possibilities have never been tried. David's account in terms of historical accidents seems much more plausible than an *a priori* assumption of the efficiency of markets. Be this as it may, my point in this essay is not to decide whether path dependence implies market inefficiency. Without doubt economists will debate these issues for some time. My argument is that the positive feedback models provide a plausible account of how historical contingency operates in technology markets.[4] Furthermore, as will be discussed below, it casts efficiency in a new light. In what follows I want to grant the plausibility of positive feedback models and discuss the implications for the concerns of praxiologists.

4. The Three E's

Practical philosophy, or the philosophy of human action, is concerned with the three E's: effectiveness, efficiency and ethics. Wojciech Gasparski (2002, p. 121) states,

> The concept of practicality – as defined within the philosophy of practicality – applies to the activity in question. According to which practicality is defined by the three "E's," the first two being praxiological values, and the third "E" comprising ethical values. … Effectiveness and efficiency are economic values. *Effectiveness* refers to the result of the action, or its effect. *Efficiency*, on the other hand, refers to the efficacy of the action itself, or the process of change. Effectiveness and efficiency appear as values that are economic *sensu stricte* when they can be measured in monetary units, and they are economic values *sensu lato* when they are of qualitative nature. Contrary to economic values ethical values are qualitative.

His article goes on to examine the relationships between the praxiological and ethical values in business practice. In conclusion, he (Gasparski, 2002, p. 134–5) remarks,

> Particular teams, particular organizations, professional societies and corporations have different ways of formulating definitions of the values they accept. *Organisation and Human Systems Development*… [Gellermann, Frankel & Ladenson 1990], which is a kind of super-code, is an example of an extensively developed collection of norms. In this super-code… we can find the praxiological values of effectiveness, efficiency and alignment. Effectiveness refers to achieving desired results (objectives, visions, missions, etc.), efficiency refers to the relation between the desired results and costs, while alignment refers to the degree to which the organisation's desired objectives… are coordinated with the expectations of the organization's members […] The super-code also includes basic ethical principles, moral norms and ideals as well as definition of conditions for deviation from the norms.

Thus, an *effective* action is one that brings about the desired consequence. An *efficient* action is one that brings about the desired consequence at the lowest cost. These two praxiological concerns can be distinguished from ethical concerns: an *ethical* action is good one. (See Gasparski, 2002, p. 122; cf. Auspitz, 1992, passim, esp. pp. 434 f.) Of course, how one defines goodness is subject to controversy.[5] I do not seek to explore this controversy here. Instead I want to make four sets of observations about effectiveness and efficiency, returning to ethical considerations at the end of the paper.

First, in *principle* of course, the distinction between effectiveness and efficiency can be made. Many problems can be solved with infinite resources. And even with available resources we could, for example, feed the world. At present, starvation is a problem of efficient distribution, rather than production, of food. In practice, however, questions of effectiveness and efficiency are intertwined. For example, the Superconducting Super Collider (SSC) may be an effective device for deciding the question of the existence of the Higgs Boson (Lederman, 1993), but the US legislature ceased funding its construction on the basis of its high cost. Of course the decision to kill the SSC was political. But such politico-economic decisions are part and parcel of the practice of science and the development of technology (Latour, 1987; Rouse, 1987). This only shows that politics plays a critical role in what counts as economically viable. And the point remains: inefficient proposals, i.e. those that are not considered cost effective, are not effective. In *practice*, the line between effectiveness and efficiency is blurry.

A second issue arises from casting efficiency in purely economic terms. The question of efficiency always presupposes value judgment. Efficiency calculations require the specification of parameters over which efficiency is determined. With respect to practical examples, we cannot discuss efficiency in general. Take for example, efficiency of an automobile. One clear notion of efficiency is fuel economy. Once efficiency is defined *as* fuel economy it can be objectively measured, and the efficiency of various models can be compared. But why pick fuel economy? Perhaps for economical reasons: a fuel-efficient car has lower fuel costs. Of course there are other costs involved, the up-front cost of the car, maintenance, etc. Perhaps a life cycle costing of the car would reveal the fuel-efficient car to be the most economical car. But even this does not settle the issue. Future costs are discounted and since fuel cost occurs throughout the life of the car they are discounted whilst the initial outlay is not. Perhaps cash flow considerations dictate that one choose a cheaper car that has lower fuel economy. On the other hand, one may have an ethical commitment to the preservation of the planet. *Ceteris paribus* a fuel-efficient car pollutes less. Thus the ethical commitment to the environment may outweigh the financial component of my decision to choose a fuel-efficient car. Perhaps one day these considerations will be internalised into the economy, but at present they are externalities.

This discussion is not a deconstruction meant to trivialise the issue of efficiency. In fact it is just the opposite: efficiency considerations always involve a complex network of value judgment, far from trivial. These judgments involve which parameters across which efficiency is defined, as well as the structure of the tradeoffs we make across various efficiency considerations. Of course the economic dimensions are crucial. In the human realm, the lack of economic viability is death. We considered just one dimension of efficiency, fuel economy, for the automobile. There are innumerable others, for instance the best acceleration per unit cost. But economic considerations are not immutable. Regulation and policy influence what is economic. These regulations and policies are negotiated. As von Weizsäcker et al. (1997) point out changes in regulation can increase the ecological sustainability of our economic practices. Value judgments, including ethical ones will figure in the negotiation of the economic parameters.

The third concern is due to the fact that efficiency arguments in economics have traditionally been couched in terms of optimisation. Efficient economic processes yield maximum output with minimal cost. In ideal markets profit is minimised whilst Pareto optimality is achieved in the long run. Treating efficiency in terms of optimisation also appears in the praxiology literature (Lange, 1963, 192 cited in Gasparski, 1992 p. 455). Many commentators have pointed out the deficiencies of this account. Herbert Simon (1992, p. 40) suggests that a satisficing model provides a more realistic approach to economics. In positive feedback economies technologies are often locked in before optimality can become an issue. Without a clear project brief engineers often do not even know what the relevant parameters for optimisation could be. Sholes was trying to make a particular (commercially unviable, as it turned out) machine work. But in fact he ended up setting the standard for the "universal" arrangement for all alphanumeric keyboards. The "problem" of what keyboard layout to adopt for the PC was "solved" a century before the PC was even conceived as a possibility! The QWERTY keyboard is an effective, and indeed satisfactory, device for entering alphanumeric data into a computer, even if it is far from optimal.

Kenneth Boulding (1992) advocates an evolutionary approach to economics.[6] This is certainly the right analogy for the development of technology. Technologies and the techniques we use to manipulate them, as we will see in what follows, evolve in the same piecemeal fashion, as do organisms. Organisms are satisficers, not optimisers. The unfit perish, but the survivors are only better than their competitors, not necessarily optimal in any dimension. And so too with technology. Often far from optimal technologies survive, not due to their efficiency, but because they are so deeply entrenched (locked in) to processes that they are impossible to change.

A final consideration that must be addressed when evaluating the effectiveness and efficiency of action concerns the contexts within which the action is taken. Individuals perform actions, but they are embedded in a variety of contexts. Some of these contexts form a hierarchy. For instance, individuals act in groups (teams, organisations, communities, etc.) and these groups function within larger social structures. These hierarchies contextualise the actions of individuals. In particular what is efficient, or reasonable, for an individual to do in isolation may not be so within specific contexts. Other contexts are not part of a social hierarchy. These contexts include culture, tradition and technology. A systematic treatment of these contexts would be useful, but is beyond the scope of this essay. The role of these contexts however is implicit in the following discussion, which is a schematic exploration path dependence and action.

5. Path Dependence and Action

Economist Bart Nooteboom has applied the principles of path dependence to the process of knowledge acquisition and use within business firms. His thesis is that knowledge acquisition, and hence the knowledge individuals possess, is path-dependent:

> Higher level, complex concepts, including categories of perception, interpretation and evaluation, are formed from modular elements that in turn develop from action, which is interaction with the physical and social environment. Hence our ability to perceive and to learn, in the sense of making new cognitive functions by reconnecting parts from existing ones, depends on past development. Hence, people will perceive, interpret and evaluate differently to the extent that they have developed in different contexts, along different paths (Nooteboom, 1997, p. 68).

Furthermore, in Nooteboom's (1997, p. 68ff.) view this path dependence applies equally to cognition in social organisations. Nooteboom's work, like praxiology, focuses on human action. In particular, he sees an intimate relationship between action and cognition. In providing a cognitive framework for his model of knowledge acquisition he aligns himself with a "school" of cognitive science he calls "situated action" (SA). In Nooteboom's (1997, p. 61) words,

> According to SA the formation of patterns of neural activity is so much a matter of interaction with action that the logical sequence is in doubt. Every action to some extent modifies the patterns of neural activity, to such an extent that one could just as well say that action precedes neural patterning. It is meaningless to conceptualise the latter as being independent of context... Learning and action go together in the construction of neural patterns of activity.

Nooteboom provides no references to members of the SA school. Nevertheless, Edwin Hutchins (1995) has applied the principles of what he terms "situated cognition" to the human activity of navigation in his *Cognition in the Wild*. Although Hutchins does not speak of "path dependence" his example provides a fruitful case study of this concept.

Traces of the path dependence of the knowledge within navigation are embedded in nomenclature. The device for measuring the speed of a vessel is the "log." This name derives from the practice of throwing a piece of wood overboard and timing how long it takes the vessel to move away from it. "Starboard" and "port" derive from a time when vessels were steered with an oar ("steering board") attached to the (right looking forward) side, which had implications for how the vessels were docked in port. Hutchins points out that there are always several ways to solve the same problem in navigation. For instance, dead reckoning[7] computations of speed need to be carried out continually. Knowing the algebraic expression $D=RT$ allows one to compute speed (R) from the knowledge of the distance covered (D) since the last computation (T) (Hutchins, 1995 p. 145ff.). However, the computation takes many steps, each prone to error. Whilst algebra provides a general abstract tool that can solve a wide-range of problems, in practice, such a device as a nomogram dedicated to solving the problem at hand is preferable.

An even more direct way of calculating speed is the three-minute rule. It is based on the relationships between time, speed and distance and the nautical mile and the yard. The nautical mile is interesting enough in itself. Landlubbers have probably wondered why the nautical mile is slightly longer than the statute mile that Americans still use to measure distance over land. I am unsure of the origins of the statute mile, but the nautical mile is certainly not arbitrarily defined. A nautical mile equals one sixtieth of a degree latitude. The Mercator projection used in constructing nautical charts is designed to minimise distortion of these distances when the globe is projected onto a flat surface. Hence the latitude scale running along the side of the chart serves as a distance scale as well. It turns out that one nautical mile is equal to 2,000 yards. We all know that there is 60 minutes in an hour. So if we measure the distance in yards a boat has traveled for 3 minutes it is very easy to see the average speed in knots (1 knot = 1 nautical mile per hour) during the interval. Three minutes is 1/20 of an hour so the speed in knots equals the distance in yards divided by 100. So you just knock off the last two zeros. For instance, say the boat has traveled 500 yards in three minutes then the speed is 5 knots.

This is very handy for dead reckoning. One needs to know the position, speed and direction the vessel is traveling in order to predict where it will be in the future. Position fixes are continually plotted by the navigator. The direction can be anticipated by the state of the helm, the current, etc. The speed must be calculated from the difference of the prior two position fixes. When near land,[8] the convention is to take position fixes at three-minute intervals. Hence the "three-minute rule" makes the determination of speed practically automatic.

What a fuller treatment of Hutchin's subject than we have room for here would reveal is a rich history of the path dependent development of navigation. Sea craft emerged independently in numerous cultural contexts as transport technologies thousands of years ago. Technologies differed across cultures. Whilst Europeans favoured wooden monohulls, the Polynesians developed catamarans. Both used sails, although there was a wide diversity even in Europe in their design. Even today the virtues of such ancient designs as gaff and junk rigs are extolled. The enormous variety of sailing vessels evolved over human history. To be utilised effectively the techniques of navigation were developed alongside. As Kotarbiński (1966, p. 202) points out, the complexity of these tasks increase as the aims become more ambitious:

> But let us revert to statement that quantity changes into quality. Many examples of such progress are supplied by increases in material, in apparatus, in the numbers of staff, etc. One can swim across a narrow river, but has to use a boat to in order to cross a much wider one; when it comes to oceans, then only ships can cross them. The differences in distances have undoubtedly been reflected in the advances in the art of navigation. And the difference between a small boat and a large ship is not only a difference in dimensions, but also of motive power (hand operated oars versus screws driven by steam or combustion engines). Moreover, the abilities needed to row are different from the abilities needed to operate a ship's engines. And that is the point I wish to emphasize. In general: an increase in material results is an increase in, and a growing complexity of, apparatus, thus forcing the agents to innovations and shifting the importance of abilities of the agents concerned towards ever increasing intellectualization and the domination of intelligence and knowledge over predominantly manual abilities.[9]

The techniques of navigation evolved independently of boat design over the same time period, as numerous souls toiled to solve the practical problems of arriving intact. The popular books *Longitude* and *Mercator* tell the stories of how two such innovations were developed in the West. Hutchins (1995, ch. 2) provides an elegant account using the principles of situated cognition to make sense of the Micronesians solution to the problem of navigation. I'm sure I do not need to inform my readers of the crucial role that seamanship played in world economics and politics.

My point here is not to recount that history, but to show how the concepts of path dependence and lock in illuminate human action within the context of history. The navigational practices we have today are the result of path dependent processes. Each procedure has a rationale that makes sense within its historical context. The definition of the nautical mile is arbitrary insofar as it could have been defined, with equal convenience, otherwise. The distance that makes up a yard was developed in a different context, and adopted in the marine context. Understanding the origins of these units is just like understanding the origins of the QWERTY keyboard. But once they are defined as they are the three-minute rule follows mathematically. Thus the procedure of the three-minute fix cycle makes sense. At various points in history practical problems were solved in a particular way; once solved they become locked in and form constraints on future solutions to future problems. What look like, out of context, to be an arbitrary set of conventions make sense as a whole. Practices evolve as a series of progressive constraint, each historical moment in the process limiting certain possibilities whilst enabling others.[10] Once evolved the whole system hangs together. A change in one element has implications throughout.

What can we learn from the path dependent nature of technological evolution? The first lesson is that there is no inevitability in the technology we possess. If one or two minor historical contingencies had gone differently I could have been typing on an "Ideal" DHIATENSOR keyboard. Furthermore, whilst the technologies we possess may be satisfactory, there is no reason to expect optimal efficiency from technology locked in through path dependent processes. Understanding both the advantages and the deficiencies of a given technology requires appreciation of the historical contingencies involved in its development. Much of the "rationality" of apparently archaic devices and techniques is hidden, remaining embedded in a tangled network of historically evolved practices. Improving technology and techniques is both possible and desirable, but the path to progress is tortuous with numerous dead-end sidetracks. *A fortiori* reforms grounded only in general principles of management, ignoring the specific details of the practice in question are doomed to failure. Insiders often cannot articulate the reasons they do what they do, even when they slavishly follow what appear to outsiders outmoded procedures.

Thus the improvement of technology and the techniques that utilise it becomes a subtle – and messy – affair. Over half a century ago Tadeusz Kotarbiñski (1966, p. 197) wrote,

> That is often precisely the error into which fall those slavish followers who imitate their master not in what is essential [...], but blindly follow his precepts under circumstances in which he could act otherwise [...] The slavish followers are routinists, since routinism consists in replacing creative innovation by the automatic replication of previous actions. Thus, standard solutions, which are one of the factors in progress in building, may under certain circumstances become an obstacle to hitting on the most rational solutions.

Oskar Lange (1963, p. 102–103; cited in Gasparski, 1992, p. 451) pursues this theme in terms of the distinction between the essential and the accidental:

> Abstraction plays a particularly important part in political economy. This is because the economic process is very complex. The economic process is a set of highly diverse actions repeated many times by a great number (sometimes reaching millions) of people, performing various activities in varying conditions. [...] These activities affect man's material environment and are themselves affected by it; [...] Abstraction consists in the mental elimination of everything which is causal or accidental, i.e. occurring only occasionally, and in singling out the event which constantly under given conditions reappears, which constantly repeats itself, i.e. is essential, or necessary. [...] The more complicated the subject of investigation, the more numerous the accidental elements which intertwine with the essential elements, the greater the number of essential elements in the process under examination the greater the importance of abstraction in research.

Lange is certainly sensitive to the subtleties involved, but the understanding of path dependent processes allow us to make an important point. We can clearly distinguish between essential and accidental features of technology in the abstract. In the course of evolution of technology however, we see historical accidents transformed into features that are practically essential. The process by which we inherited QWERTY is a path dependent series of accidents, but once locked in it is for practical purposes as essential as if it had been a gift from the gods. In understanding path dependent technological evolution historical contingencies often play an *essential* role, and hence cannot be ignored.

The explosive growth in personal computing technology over the past three decades is an illustrative case of the messiness of technological advance. The exchange value of PC's cannot be denied. According to Edward Tenner (1991, p. 103), "billions of dollars" were spent on "microcomputers [to replace] batch processing and mainframes". But the utility is more difficult to evaluate. Tenner (1991, p. 103–104), continues,

> If this unprecedented power had performed as advertised, productivity in services should have soared. Instead it increased only 1.3 percent a year between 1982 and 1986. In 1989–90 it grew by only 0.5 percent. The largely precomputer postwar average growth had been fully 2.3 percent.

Tenner explores in great detail the causes for the lack of productivity gains, the result of what he calls "revenge effects" in his *Why Things Bite Back* (and I invite the reader to peruse his sprightly account), but a few moments reflection on our everyday practice yields some insight into the problem. How much time do you spend on composing, reading, replying to, sorting and deleting email every day? How much of this communication is essential, or even useful? How much of this communication could be handled more efficiently through face-to-face or telephone communication? How much time is spent keeping your computer free of viruses? How much time do professionals spend typing, formatting, printing, photocopying their own documents? How many times have you reprinted an entire document when a minor error is found? How many people actually become proficient in using a software package before it becomes obsolete? How long do you idly wait for your computer to boot up or perform some task? How much time do you spend asking colleagues (who are not computer professionals) how to perform a task on your computer? How much of your colleagues' time does this take away from their normal duties? Tenner (1996, pp. 253–-254) reports the results of a study into that last question:

> Peer support often turns out to be a rearranging effect. We save time by having computers accomplish things for us, but before they can do them we take time from our colleagues. [...] Most amateur [computer] gurus do not mind interrupting what they are supposed to be doing. [...] Even if they love their work, they probably find peer help psychologically rewarding and politically valuable in accumulating allies and favors owed. [...] Since some peer helpers are world-

> class scientists, engineers and other professionals, the information services department is getting a bargain – and the organization is losing part of the high-priced time it is paying for. Yet without them, as William M Bulkeley wrote in the *Wall Street Journal*, "millions of U.S. workers might turn their PCs into expensive hat racks."

Tenner (1996, pp. 253–254) cites a study by Nolan, Norton & Co. that estimates the cost of such peer support to be $ 6 000 to $ 15 000 per year per computer, or about three times as much as "the cost of equipment, supplies and in-house technical assistance." Eventually we will learn to live more efficiently within a computerised workplace, but this will be a slow, costly process. Each solution we adopt will lock us in to unforeseen new inefficiencies forming part of our work environment. This process is clumsy, but typical of technological progress.

So, understanding the reasons behind the technologies we end up with requires understanding the historical details of the processes that led to them. The positive feedback cycles involved in such processes implies that the effects of minute details can become amplified to produce significant effects. This leads to a dilemma when taking action in such situations: before lock in occurs the most delicate action may cause the change one desires. However, at the point before lock in it is impossible to predict what the outcome of any action will be. Of course, after lock in the outcome is clear, but then it is very difficult to do anything about it. This feature allows us to read Kotarbiñski's (1965, p. 163ff.) strategies of *fait accompli* and deliberate delay in a new light: gamblers will prefer the former, betting on their preferred outcome being locked in. The latter strategy is more conservative: learn how to profit from the locked-in result. The former strategy has the highest potential for a large payoff, but also the highest likelihood of failure.

Successful action requires predicting outcomes. But there is even a problem for understanding *a posteriori* the evolution of technology. Whilst the effects of technological decision making may be locked in for posterity, all traces of the microscopic origins behind the effects may be lost. For instance, perhaps a crucial aspect of the real inspiration for the QWERTY keyboard came from a breakfast conversation with Mrs. Sholes on the morning of 17 September 1865. Of course, the likelihood of such a detail being historically recorded is low, especially given that at the time the auspiciousness of hypothetical event would not have been appreciated. Don McCloskey (1990, p. 86) makes the same point with respect to historical narrative:

> Little events can have big consequences in some parts of the history. The parts are described by models that are nonlinear in the events and whose consequences feed on themselves. That is, a little event put into one of the equations yields a large consequence, which is fed back as input. "Nothing succeeds like success" is such a model [...].

McCloskey (1990, p. 88) summarises his account of nonlinear feedback thus: "But if, as... many historians have persuasively argued, the correct model for [a particular historical episode]... are models of nonlinear feedback, then the story becomes unmanageable, untellable. It is a paradox." Positive feedback models of technological development imply that the microscopic details matter, making the job of the historian crucial but difficult, if not in the end impossible.

Being effective requires the ability to predict the outcome of our actions. In some cases understanding the context of our actions may increase the odds of accurate prediction. Where positive feedback dominates the dynamics of the situation accurate long-term prediction is impossible. One might respond to this fact with despair. I have seen the following prayer reprinted on items sold in gift shops across America: "God grant me the strength to change what I can, the grace to accept what I cannot, and the wisdom to know the difference." The ability to identify sources of positive feedback and hence recognise the epistemic context of ones situation can provide such wisdom.

Nevertheless, a little sophisticated reflection on nonlinear models can help us improve the situation further. The key to effective action in positive feedback contexts is discerning the proper scale of understanding and intervention. This science is in its infancy, but a hint as to the path to success is provided in the work of biomedical researcher Ary Goldberger. His problem is understanding the nonlinear dynamics of heart disease. At a fine level of detail human heart rhythms are impossible to predict. But Goldberger's (1997, p. 548) research indicates that there are differences between the patterns exhibited by healthy and diseased human heartbeats detectable through global analysis of the complexity of ECG signals. At this point the clinical implications of this work are unclear, but we can be sure about this: finding the right scale for intervention is crucial. And perhaps we can apply this insight to technological development. We may not be able to control the details, but we may be able to place technological markets on a more efficacious trajectory if we can identify the right scale of intervention.[11]

Nooteboom (1997, pp. 63–64) provides a further clue. He advocates modeling our organisations on our minds and language:

> Scripts can be represented as lines, indicating the action sequence, with nodes, which function as variables that allow for the substitution by diverse items (such as tools, materials, and the like). This seems to be the appropriate manner to conceptualise reference: a fitting into scripts... This implies allowance for dubious substitutions, which may, when successful, turn out to be innovations... On the other hand, for a minimum of stability in practice, some obstacle to unorthodoxy must be present. This brings us back to an argument from Kuhn: some conservatism in maintaining incumbent practice is rational... But too much conservatism inhibits adaptation to changed conditions. Survival requires a balance of, on the one hand, routine, habit, conservatism, continuity, and on the other hand, adaptability, innovation, shift. What the right balance is depends on the volitility of the environment. Our minds and language are masters at this. How about our organisations?

Furthermore, it is the path dependence of the knowledge possessed by individuals that can provide the diversity required for fueling innovation. As was argued above the unique path of each individuals development that lends diversity of perspectives. Functioning within an organisation individuals are governed both by organisational and individual intersests:

> As argued by Smith Ring and Van de Ven (1994), people act privately, 'qua persona', and according to organisational roles assigned to them. Alignment between the two can be a problem, but this duality of conduct can also be an advantage. [...] the diversity of personal perception, understanding and evaluation can act as a creative force, to yield novel perceptions, ideas or directions for an organisation (Nooteboom, 1997, p. 69).

So from one perspective, positive feedback, along with the path dependence and lock in it implies is problematic. It makes our technological systems unwieldy, difficult to predict and control. But from another perspective, positive feedback provides the dynamism required to adapt and to innovate in a changing environment.

6. Conclusion: Path Dependence, Ethics, and Values

In the foregoing section I focused mainly on the implications for efficiency of the path dependence of technological development. In this final section, I would like to fulfil my earlier promise to address the "third E" of practical philosophy, ethics. The remarks here will be brief, but I think they follow in a straightforward way from what has preceded.

Finally, I will argue that the existence of positive feedback in technology markets motivates a careful attention to the fundamental values underlying technological decision making.

Allen Buchanan (1985, pp 51ff.) discusses the "argument from desert" used to defend the free market on moral grounds.[12] As Buchanan points out, there are really two forms of the argument, "but both assume that the market *does* distribute wealth according to desert and both assume that wealth *should* be distributed according to desert." Whilst Buchanan does a fine job of rebutting both arguments, what I want to add here is that positive feedback in markets rewards a little effort put in the right place at the right time. However, since it is impossible to know *a priori* where to put in the effort, success is largely a matter of luck. Of course, being in many places much of the time increases ones chances of being in the right place at the right time, and being in many places involves hard work. Nevertheless, hard work is only a necessary condition for such a strategy to work. Most entrepreneurs, even many with good ideas, fail. Hence the very nature of positive feedback markets result in reward and effort being out of proportion.

So, reward for success as well as responsibility for failure do not always rest primarily with the developers of technology. The amplification due to positive feedback of small events into huge successes (and huge failures) can provide the answer to McCloskey's American question: "If you're so smart, why aren't you rich?" It applies equally to the clerks retort: "If you're so rich, why aren't you smart?" (McCloskey, 1990, p. 128) Thus the situation is much as it always has been in capital markets; however, the concept of positive feedback explains why a huge gap can open up between the winners and losers when only a hair's breadth separates the competitors at the outset. This phenomenon is endemic to the market, and whether one deserves the fruits of such success or not is independent of the level of reward.

Finally, the subtle nature of the dynamics of technological systems has important implications for the praxiology of technology. As argued above, path dependent processes evolving over extended periods of time yield complex technological organisational networks. Whilst these networks are the result of human activity, often no particular individual has a grasp of even a large part of the network. Furthermore, the decisions made to construct the networks are constrained by previous decisions, technological capability, as well as social and cultural factors. Hence many of the actions taken are directed at preserving the network rather than accomplishing the tasks the network was designed to achieve. These terms provide one way of making sense of the "autonomous technology" thesis.[13] We do not have to be committed to technology literally having a mind of its own to see that a threshold can be crossed whereby most of our effort goes into constructing and maintaining the technological infrastructure supporting the activities that are our true aims. Positive feedback makes a contribution to driving this dynamic. This leads to a situation whereby two kinds of concerns pull in opposite directions.

On one hand, it is prudent to apply the precautionary principle when improving efficiency in high technology systems. The complexity of these systems makes ignorance endemic. Furthermore, positive feedback implies that the magnitude of effects can be disproportional to small structural changes that brought them about. And often it is unclear where within the structure positive feedback lurks. The issue is compounded by the effect redundancy has on efficiency. Nooteboom (1997, p. 66) makes the following point in his discussion of neural networks:

> Note that there is a trade-off between specialisation and functional redundancy: specialisation (assignment of functions to fixed units) is more efficient, due to habituation, but less flexible than functional redundancy (ability to perform different functions within one unit), and requires greater overhead and more hierarchy for communication and coordination.

Thus, systems without redundancy require more resources (and I would add greater sophistication) for coordination. In general though, redundancy is the enemy of efficiency; specialisation and streamlining providing the main paths to reducing costs whilst maintaining productivity. Nevertheless, systems streamlined in such a way can become brittle; redundancy is necessary for reliability. This applies equally to social organisations as it does to neural networks. Thus short-term efficiency gains can in the long run bring about systemic failure, or even collapse.

On the other hand, we must always be on guard against "becoming slaves to our machines" in the sense that we invest more effort into maintaining them than we invest in accomplishing the tasks for which the technology was designed. As Kotarbiński (1966, quoted above) observes, the more ambitious our projects are in terms of material results, the more complex our technologies (and I would add the organisations that control them) become. Hence, increasingly we rely on others to produce the tools we use. This decreases the control that individual users have over technology whilst increasing the odds that the technologies will fail to fall in line with the value priorities of the user. Just a small example from the realm of personal computers: My word processor performs numerous functions that I do not need or desire. For example, there are 71 fonts available in my version of *MS Word 95* in sizes ranging from 8 to 72 points. Despite this fact, everyone I know uses Times New Roman, 11 point! And of course we all pay for lines of code we have no use for when we purchase a PC. This code may increase user-friendliness. However, user-friendliness is usually a marketing ploy, attracting new users. And these features often stand in the way of efficient utilisation of the PC for experienced users. Very few people have the time and expertise to design their own software, and even if they did it wouldn't help. This is one of the lessons of positive feedback markets: the need to share information is a major source of positive feedback in the marketplace. Hence one is stuck, like everyone else, with what the major manufacturers provide. And of course, within reason, their priorities often take precedence.[14]

Nevertheless, individuals are often complicit in the process. They acquire things they do not need, and can lose sight of their own goals. I am not here advocating a Luddite thesis. Whilst positive feedback models of technological development can explain how technology can take on a life of its own, I do not advocate a life free of high technology. Nevertheless, I do think that the world is in need of some careful critical analysis of the values underlying our technological priorities.

Notes

1. Positive feedback can work at other spatio-temporal scales, and outside technology markets; Kenneth Boulding (1992, p. 53) describes how positive feedback cycles can lead to depression.
2. Paul Ormerod (1998, pp. 22–24) points out that Arthur's model is a simplified special case of the "ant models" that form the basis of his book. Whilst Ormerod's scenarios are intriguing, I can make all the points I would like to in this paper with Arthur's models, which are simpler to understand.
3. And I suppose that this point generalises to all languages using the Roman alphabet.
4. Some quasi-markets, e.g. academic and scientific reward are subject to positive feedback and thus can be plausibly described by such models.
5. In basic terms a consequentialist defines a good action as one that brings about a desirable result, a deontologist will have principles that no action ought to violate, independent of its consequences (for an extended discussion of the extent to which Kant is a deontologist see Atwell, 1986) and a virtue theorist judges goodness in terms of conformity to certain virtues.

6. Evolutionary economics has an extensive pedigree. See Foster 1987 for a good overview supplying the historical context.
7. By the way "dead reckoning" has nothing to do with death. It is shorthand for "deductive reckoning" = "ded. reckoning" which when spoken sounds like "dead reckoning"!
8. Land is the main source of danger for a boat. No vessel has ever run aground in the open sea!
9. There are two caveats here. 1. It must be emphasised that the "intellectualization" referred to here is based on "situated" rather than "abstract" cognition. 2. Human beings regularly outstrip their capacities to intelligently guide their actions. E.g. oceans were crossed long before they could be reliably navigated; whilst Columbus may have "discovered" America, he set out to find a Westerly route to India!
10. The first time I encountered this account of practice was in Dyke (1988). Chapter 6 provides an account of bridge conventions in these terms without reference to "path dependency".
11. Paul Ormerod (1998) provides numerous examples of this point in socio-economic contexts.
12. Buchanan refers the reader to J. Feinberg, "Justice and Personal Desert" in J. Feinberg (ed.), *Doing and Deserving*, (Princeton: University Press, 1970, pp. 55–87).
13. For an extended discussion see Winner 1977.
14. This analysis could be applied to the capitalist system itself. In fact, positive feedback inherent in the free market system itself is partially driving the "capitalist threat" (see Soros, 1997; Zsolnai, 2002) currently coming to head.

References

Arthur, W.B. 1994. *Increasing Returns and Path Dependence in the Economy*. Michigan: University Press.

Atwell, J.E. 1986. *Ends and Principles in Kant's Moral Thought*. Dordrecht: M. Nijhoff.

Auspitz, J.L. 1992. "A Note on the Coherence of Praxiology and Ethics in the Philosophy of Tadeusz Kotarbiñski" in Auspitz et al. 1992.

Auspitz, J.L., Gasparski, W.W., Mlicki, M.K. and Szaniawski, K. (eds.). 1992. *Praxiologies and the Philosophy of Economics*. New Brunswick: Transaction Publishers.

Boulding, K.E. 1992. "Appropriate methodologies for the Study of the Economy." in Auspitz et al. 1992. pp. 43–60.

Buchanan, A. 1985. *Ethics, Efficiency, and the Market*. Totowa: Rowman & Allanheld.

David, P.A. 1986. "Understanding the Economics of QWERTY: the Necessity of History." in W. N. Parker (ed.) *Economic History and the Modern Economist*. Oxford: Basil Blackwell.

Dyke, C. 1988. *The Evolutionary Dynamics of Complex Systems: A Study in Biosocial Complexity*. New York: Oxford University Press.

Foster, J. 1987. *Evolutionary Macroeconomics*. London: Allen & Unwin.

Gasparski, W. 1992. "Oskar Lange's Considerations on Interrelations Between Praxiology, Cybernetics, and Economics." in Auspitz et al. 1992. pp. 449–464.

Gasparski, W. 2002. "Effectiveness, Efficiency, and Ethicality in Business and Management.." in Zsolnai 2002. pp. 117–136.

Goldberger. A.L. 1997. "Fractal Variability versus Pathologic Periodicity: Complexity Loss and Stereotypy in Disease." *Perspectives in Biology and Medicine*. *40*: 4 (Summer). pp.543–561.

Hutchins, E. 1995. *Cognition in the Wild*. Cambridge: MIT Press.

Kotarbiñski, T. 1966. *Praxiology: An Introduction to the Sciences of Effective Action.* O. Wojtasiewicz (tr.) Oxford: Pergamon.

Latour, B. 1987. *Science in Action: How to Follow Scientists and Engineers through Society.* Cambridge: Harvard University Press.

Lederman, L. 1993. *The God Particle.* New York: Delta.

Liebowitz, S. and S. E. Margolis. 1996. "Typing Errors", *Reason.* June.

McCloskey. D.N. 1990. *If You're So Smart: The Narrative of Economic Expertise.* Chicago: University Press.

Nooteboom, B. 1997. "Path Dependence of Knowledge: Implications for the Theory of the Firm." in L. Magnusson and J. Ottosson (eds.) *Evolutionary Economics and Path Dependence.* Cheltenham: Edward Elgar.

Ormerod, P. 1998. *Butterfly Economics.* New York: Pantheon.

Rouse, J. 1987. *Knowledge and Power: Toward a Political Philosophy of Science.* Ithaca: Cornell University Press.

Simon, H.A. 1992. "Methodological Foundations of Economics." in Auspitz et al. 1992. pp. 25–42.

Soros, G. 1997. "The Capitalist Threat." *Atlantic Monthly 279*: 2 (February). p. 45.

Tenner, E. 1991. "Revenge Theory." Reprinted in M. E. Winston and R. D. Edelbach (eds.) 2000 *Society, Ethics, and Technology.* Belmont: Wadsworth.

Tenner, E. 1996. *Why Things Bite Back.* New York: Vintage.

von Weizsäcker, E., Lovins, A.B. and Lovins, L.H. 1997. *Factor Four: Doubling Wealth–Halving Resource Use.* St. Leonards: Allen & Unwin.

Winner, L. 1977. *Autonomous Technology.* Cambridge: MIT Press.

Zsolnai, L. (ed.). 2002. *Ethics and the Future of Capitalism.* New Brunswick: Transaction Publishers.

PART TWO
Action and Innovation

Praxiology and the Effectiveness of Information and Communication Technologies

Krzysztof Klincewicz
School of Management
Warsaw University
Poland

1. Introduction

The effectiveness of Information and Communication Technologies (ICT) has recently become a widely-discussed topic among practitioners and researchers. Current debates do not however attach adequate importance to the actual nature of the means-end relationship, which is described by the very notion of effectiveness, taking instead for granted available means (interpreted as static, well-defined technological objects) and ends (unanimously presented as efficiency improvement). The article will review the discussions, explaining sources of incommensurability among diverse postulates, and indicating underlying mechanisms, responsible for interpretations of the effectiveness within the praxiological framework.

Technology has traditionally been regarded as "hardware" in organizations: physical machines and equipment used to support manufacturing processes (Hatch 2002: 158). First empirical studies interpreted it as determining other organizational variables, while early research on organizational innovations distinguished technical innovations (involving technology) from administrative innovations (related to social aspects of organizations) (Daft 1978: 197–198). However, the rapid development of Information and Communication Technologies (ICT, also referred to as "new technologies") in the recent 20 years changed the research perspective, as these new technologies possess radically different characteristics, blending the technical and social realms, no longer determining, but rather facilitating and flexibly adjusting to organizational processes (Kieser, 1990). ICT transform advanced technical knowledge into hardware and software solutions, supporting human interactions and data processing, present nowadays in practically all aspects of everyday life.

The problem of ICT effectiveness is relevant and important for practitioners due to the substantial scale of related expenditures, as well as the critical dependence of organizations on their infrastructure. At the same time, there are many opinions about limited effectiveness or even counter-effectiveness of these investments, which started with the formulation of the so-called Solow paradox in 1987, and dominated the management literature in years 2000–2001, during the high-tech market depression. Moreover, postmodern interpretations of technology as text (loosing its tangible dimensions) are gaining on popularity, radically changing the debates about ICT by rendering the sole notion of effectiveness irrelevant (Joerges, Czarniawska 1998).

Praxiology seems to offer a suitable perspective on the problem of effectiveness due to its focus on means-end relationships. The Misesian analytical framework (Mises 1996: 92–98) emphasizes the multiplicity of possible ends, directing human action, and extending beyond merely financial objectives. It also proposes to look at available objects (technologies) through the lenses of their relevance for attaining given ends by acting individuals, so that the objects can become means. The apparently intuitive method seems however to have been forgotten by many investigators – the present article will thus critically review existing research studies concerning ICT effectiveness, further referring to the seminal perspective on means-end relationships to propose a complementary, praxiological approach.

2. Four Views on the ICT Effectiveness

The existing literature addressing the problem of ICT effectiveness can be divided into four broad streams, based on its evaluative focus (emphasis on positive or negative effects), and assumptions about the impact of ICT on the social sub-system of organizations. The latter criterion deserves a more detailed explanation. Publications assuming that the social sub-system is not affected, are not concerned with possible modifications of organizational structures, processes or cultures. This technocratic perspective presumes that the development of technology unavoidably leads to its diffusion. An opposite orientation, presented below as "social approach", appreciates the importance of social changes, accompanying the introduction and use of new technologies, emphasizing interactions between technology and other organizational variables.

Table 1
Technocratic and social approaches to ICT effectiveness (source: own)

	Technocratic approaches	**Social approaches**
Organizational decisions	optimum decisions based on analysis and calculation	optimization not possible due to environmental uncertainty
Requirements	specific	unclear and complex
Results	measurable, objective	difficult to measure, multidimensional, subjective
Character of results	quantitative	quantitative and qualitative
Technology development	linear improvement of technical parameters	complex processes involving social change
Scope of effects	restricted to existing activities and operations	impacting other subsystems, with feedback loops
Change	predictable	revolutionary, unexpected
Solutions	standard systems	multiple, creative designs
Discourse types represented ICT ineffectiveness	ICT effectiveness	ICT counter-effectiveness social construction of ICT

As Table 1 suggests, the technocratic approach is rooted in positivist assumptions about the organizational reality, according to which decision makers in organizations can clearly specify their requirements, so that the effectiveness of technologies could easily be measured. The requirements and results are quantified in financial terms of cost reduction, productivity, or return on investment. Technology development is further understood as linear improvement of technical parameters, making future changes predictable. The role of new technologies in this interpretation is restricted to supporting existing operations, while not inducing any social changes.

The social approach appreciates in turn reciprocal relations between human action and technology development, when the introduction of a new technology causes organizational reactions. Due to the openness of organizational systems and high degrees of environmental uncertainty, organizations are not able to precisely define objectives, and their requirements are constantly changing. Results of a technology implementation may therefore affect various domains, leading to both quantitative and qualitative change.

Table 2 outlines four distinctive discourse types, resulting from an intersection of the two variables, described in detail in the following sections.

Table 2

Four types of discourse concerning ICT effectiveness (source: own)

	positive effects	negative effects
social subsystem not affected	**ICT effectiveness (ICT as a solution)**	**ICT ineffectiveness (ICT as cost)**
social subsystem affected	**Social construction of ICT (ICT as a tool)**	**ICT counter-effectiveness (ICT as evil)**

3. Discourse Type 1: ICT Effectiveness

The core of technocratic approach is represented by this discourse type, based on implicitly assumed, positive role of technology in solving organizational problems. Traced back to Schumpeter's presentation of the role of innovations in national economies, the perspective applies a biological analogy between organisms and technologies, which have their lifecycles, undergo evolution (continuous improvement of technical parameters) and natural selection (substitution by superior technology generations). Its focus on technology development, not diffusion, led to excessive optimism about the acceptance and usefulness of any technical innovation ("pro-innovation bias", Rogers 2003: 106), ignoring the importance of people as designers and users. Technology is actually both a product of human action, as well as a force conditioning and shaping the very same actions. The humanistic paradigm of design calls for substantial user involvement in shaping the technology Gasparski 1991. The paradigm dominates the present ICT development practices, even though it does not always find its way into the academic literature. In spite of the criticism and limited relevance for actual practices, the ICT effectiveness perspective dominates the technology-related discourse. The role of technology in organizations is thus limited to efficiency improvement of existing practices and reduction of costs (cost displacement function) while ignoring possible social dynamics, establishment of entirely new practices, or changes to the ways people work and live (added value function) (Sproul, Kiesler 1994: 4).

4. Discourse Type 2: ICT Ineffectiveness

The second perspective is rooted in the same technocratic view of technology development, focusing however on the outrageous scale of business expenditures on computing solutions and disappointing results of these investments. Numerous case studies of failed projects apparently support a general conclusion that instead of the promised cost reduction, ICT do actually generate hidden costs. Countering the optimism of technology supporters, authors quote examples such as substantial investments needed to address the so-called Y2K problem (expected failure of legacy software, which would not cope with dates after December the 31st, 1999), growing security threats (linked to the popularity of the Internet) or information overflow, reducing personal productivity instead of improving decision making processes.

Robert Solow, Nobel laureate in economics, formulated in 1987 the famous Solow paradox, which assessed negatively the effects of ICT on the American economy ("You can see the computer age everywhere, but in the productivity statistics") and started an ongoing productivity debate among economists. Statistical data from the following years seemed to counter this statement (Karsenti 1999), but other economists convincingly argued that the models were distorting real impacts of technologies in many ways, e.g. by incorporating the investment in ICT by… ICT companies themselves (Gordon 1991). With the increasing complexity of solutions, "even more powerful personal computers wind up being used to type letters, and the letters are not types appreciably faster" (Triplett 1999: 315–316). Computers became solutions looking for possible problems to address, unquestionable ways to accomplish certain organizational ends, rationalized myths ("*technologizing* as the sole objective" (Faust, Bahmüller 1996: 1340)). Another widely-discussed article was published in 2003 by Harvard Business Review under the self-explaining title "IT doesn't matter" – Nicolas Carr argued that new technologies cannot offer real strategic advantages, becoming commoditized, taken-for-granted infrastructure (Carr 2003).

While offering interesting insights, the interpretation of ICT. as primarily ineffective is based on numerous misinterpretations. Automation of inefficient processes is bound to generate negative results, and excessive investments could not guarantee positive economic effects (Picot, Reichwald 1998: 187). There are nevertheless numerous examples of successful technology projects, or positive effects of ICT for entire industries, so the generalized criticism misses important aspects of the discussed problem.

Both of the discourse streams introduced so far – ICT effectiveness and ineffectiveness – represent the technocratic approach with its idealistic picture of technologies, progressing through the development of technical parameters, effective or not in inducing the desired changes in organizational performance. Within the confines of this approach, a dialogue or reconciliation of two opposing views are difficult due to contradictory values, but the problems might be overcome if the analytical perspective is extended to include social changes induced by new technologies, as postulated in the remaining types of discourse.

5. Discourse Type 3: ICT Counter-effectiveness

This type of discourse regards ICT as a source of radical social change, having primarily negative effects on users and on the society. Technical innovations do not improve organizational efficiency despite wide-spread beliefs and propaganda, reporting apparent benefits experienced by some users. Today's technologies create tomorrow's costs and unexpected risks, which in most cases are difficult to foresee when the initial implementation decisions are made. The omnipresence of ICT can induce violence (computer games), popularize pornography (Internet), abuse intellectual property (peer-to-peer computing), encourage hacking or frauds, and invade privacy. This perspective is rooted in Foucault's interpretation of knowledge as power, and technology as a rationalized mechanism of control. The apocalyptic visions of de-humanization and hidden risks of control are offered by popular literature, including classics like Boris Vian,

Gabriel Garcia Marquez or George Orwell, as well as science-fiction authors. Similar attitude is represented by journalists, looking for sensations such as Y2K problem, spread of new computer viruses, Internet addiction or the use of information networks by state intelligent agencies. Technology is presented either as created for mistaken purposes, or easily misused against the intentions of its creators – just like a hammer, originally designed to facilitate nailing, can also be used to kill people (Jonas 1996: 105–106).

The main problem of this perspective is the implicit assumption that technological development is disastrous to the human race, and ICT could not meet requirements of users, rather manipulating people (by capitalizing on their naiveté) and inducing addiction (as a form of subconscious influence or simply an escalation of necessary investments due to hidden costs).

6. Discourse Type 4: Social Construction of ICT

The last perspective emphasizes the positive role of new technologies, at the same time attaching importance to the dynamics of social change. Technologies may – but not always do – satisfy the needs of organizations, and achieved benefits are always related to the initially defined requirements and the actual patterns of use. Moreover, consequences of an implementation extend beyond the set of problems originally addressed, so organizations are faced with multi-dimensional changes, resulting both in unexpected benefits, as well as unforeseeable problems – the notion of "change" seems to capture the dynamics more adequately that the term "technological progress", which entails positive evaluation of the developments. Measurement of achieved benefits (and thus ICT effectiveness) is complicated, as the technology not only improves efficiency, but also adds value in areas difficult to specify and quantify (Anandarajan, Wen 1999: 330).

An important contribution was offered by Orlikowski, who argued that technologies link physical aspects (form and functions, determined by designers) with social dimensions (depending on the context of use). This dualism means that "technology is created and changed by human action, yet it is also used by humans to accomplish some action. (...) Once developed and deployed, technology tends to become reified and institutionalized, losing its connection to human agents that constructed it or gave it meaning, and it appears to be part of the objective, structural properties of the organization" (Orlikowski 1992: 405–406). This apparent stability does not however mean that technology becomes a deterministic power, as the entire influence of technology on human action results from rules, defined and always modifiable by users and designers (Orlikowski 1992: 411).

New technologies interact with people, who constitute their features by exploiting opportunities, offered by the technical artifacts (Orlikowski, Hoffman 1997), often not foreseen by the original designers or even contradictory to their original plans – in her work, surprisingly resembling the classical praxiological perspective on human action and its outcomes, Orlikowski outlined a concept of "technologies-in-practice" (Orlikowski 2000), opposing traditional views of reified, stable and deterministic technologies. New technologies can easily be adjusted to specific organizational requirements, but on the other hand, also their interpretations, valuation and patterns of use are shaped by individual attitudes and value systems (Combs, Knights 1992: 55), constituting "interpretative flexibility" of technologies (Bijker, Hughes, Pinch 1987). Organizational psychologist Karl Weick argues that technology is an *equivoque*: "something that admits of several possible or plausible interpretations and therefore can be esoteric, subject to misunderstandings, uncertain, complex, and recondite" (Weick 2001: 148). The same physical technology may be adopted to support both flexible forms of organizing, leading to decentralization and human networking, as well as strongly centralized processes with intrusive control mechanisms – consequently, benefits of its adoption will vary among organizations, depending on the patterns of use and assumed objectives.

ICT differ from traditional manufacturing technologies, which determined organizational structures and imposed limitations on inputs, outputs and processes. Up to the late 1970's, large mainframe computers were regarded as a limiting factor for organizational design, but modern solutions can easily be adjusted to match specific (or even changing) requirements. ICT become thus a form of underlying infrastructure, which is flexible enough not to enforce structures or processes, supporting creative design of both organizations and technological systems (Kieser 1990: 173–174). Implementation projects require ongoing interactions among designers, users technologies, guided by dedicated methodologies, reference models, configuration and modeling of processes and data structures. Technologies interact with organizations, and every technology has both efficiency effects (which motivate organizations to undertake implementation projects), as well as social system effects (often ignored, but eroding potential benefits), so that "more efficiency in one place can cause less in another" (Sproul, Kiesler 1991: 1–2). The introduction of bank cards reduced costs of banking transactions – but decreased their average values; offered convenience of use to bank customers – while also exposing them to new forms of bank frauds (Triplett 1999: 319). Popular corporate e-mail systems improve internal communications and help share information, although they may also have an adverse effect on productivity by bothering employees with unwanted messages (Sproul, Kiesler 1991).

The conclusion is that new technologies help us do things differently, supporting various alternative patterns of work. Due to the incommensurable ways of using ICT, an aggregate assessment of their effectiveness could not offer useful insights into the nature of the phenomena. More plausible would be a contingent, ideographic approach, taking into account the uniqueness of every ICT implementation, as resulting from interplay between technological and social factors, set in a certain historical context, helping understand under which conditions and for which purposes the specific solutions were implemented. Effectiveness characterizes specific means-ends relationships – but as the further discussion will demonstrate, both means (i.e. technologies and their applications), as well as ends related to ICT implementations may vary significantly, and unfortunately this praxiological reflection usually does not precede economic analyses.

As Table 3 demonstrates, the four types of discourse present incommensurable arguments, focusing on different ICT characteristics. While each of them offers insights into the nature of new technologies, the most promising contribution for praxiologists seems to be provided by the research paradigm centered on the social construction of ICT, capturing the dynamic character of technologies and the role of human action in establishing means-ends relationships.

Table 3
Characteristics of four ICT effectiveness discourse types
(source: own)

	ICT effectiveness	ICT ineffectiveness	ICT counter-effectiveness	Social construction of ICT
Interpretation of ICT	solution	cost	evil	tool
Relations between technology and needs	technologies meet demands of customers	technologies do not match the actual needs of users	technologies do not meet requirements, manipulate people and induce addition	technologies may meet requirements, depending on implementation and use
Benefits from innovations	every innovation is beneficial, as technology is good for itself	innovations are not beneficial, do not solve problems	every innovation entails risks	innovations may be beneficial, their consequences extend beyond the problems initially addressed
Progress	progress occurs through improvement of parameters of technologies; quantitative measures	technologies do not progress, but become more complicated	technological progress disastrous to the human race	progress difficult to define; multidimensional quantitative and qualitative measures
Requirements and preferences of buyers	purchasing decisions based on clear goals and preferences (parallel to the process of natural selection of technologies)	purchases result from mistaken beliefs that ICT could meet expectations	purchasing and adoption decisions result from naiveté or addiction	purchasing decisions complex, without clear goals, as all consequences difficult to foresee
Nature of decision	well-considered decision	mistake	addiction	imitation and adjustments

7. Analysis of Means: Technologies-in-Practice

The social construction of ICT perspective proposes to view technologies as fluent, allowing multiple interpretations and uses – such "technologies-in-practice" seem to constantly be "in search" of their meanings and applications. Literature on information systems corresponds to this view by emphasizing the importance of so-called "killer applications", compelling reasons to adopt certain physical technologies. Teleological views of technology as having an inherent purpose cannot thus be maintained for new technologies, continuously changing and reinventing themselves. An interesting example could be peer-to-peer technology, initially employed by multimedia file sharing services and used for illegal purposes, infringing intellectual property rights – but later exploited by established vendors to support communication crossing organizational boundaries. ICT innovations are often separated from their future applications, patented by inventors before any plausible use is identified, with opportunities for using them emerging gradually. New technologies can be both *equifinal* (with many alternative means leading to similar effects) and *multifinal* (as the same physical tools can be used to address radically different problems).

Patterns of information technology use in organizations offer additional insights into the phenomenon: the sole act of purchase (or adoption) has to be followed by implementation and assimilation (Ramiller, Swanson 2003: 8–9). Only assimilated technologies can demonstrate their usefulness, while economists interpret and analyze only the adoption data. Technologies need to actually be used – the initial expenditure does not guarantee further results, and the implementation and adoption stages reveal the interpretative flexibility and creative potential of ICT.

This does not however mean that new technologies lose their physical characteristics and become only "text", as postmodern authors would suggest (Joerges, Czarniawska 1998). The textual dimensions, linked to multiple possible interpretations and uses, are still based on certain tangible technical artifacts, and new interpretations of a technology usually require additional technical development to actually demonstrate the promised benefits, so the "text" can only "live" within the physical boundaries defined by existing technical possibilities.

The object of the effectiveness debate should be precisely defined in line with the Misesian postulates for the means-end analysis: the debate should concern individual technology applications, not technologies per se, paying adequate attention to the actual patterns of their use.

8. Analysis of Ends: Multiple Motives

Three of the previously presented perspectives on ICT effectiveness assumed that the purpose of ICT implementations is the improvement of operations efficiency and productivity. However, empirical studies revealed also other motives for technology purchases, extending beyond the economic effects (Ramiller, Swanson 2003: 10).

Early research on management information systems demonstrated that managers were not really using the delivered information as inputs to decision making processes, but rather as symbols and signals of knowledge and power (Feldman, March 1981). Modern technologies could also be implemented as "organizational facades", symbolizing innovativeness for investors, partners or customers and improving corporate image (Nystrom, Starbuck 1984).

Another type of arguments justifying purchases is related to social influence. In the case of enforced compliance, organizations do not implement certain solutions to achieve measurable benefits, but to avoid punishment from regulatory bodies, requiring them to comply with specified standards (King, Gurbaxani 1994). The mechanism of mimetic isomorphism relies in turn on heuristics facilitating reasoning processes: imitation of other organizations, without actual reflection about the adequacy of implemented solutions (DiMaggio, Powell 1983). Similarly, management fashions claim to offer "best practices", apparently easy to implement and proven by other organizations. In these cases, efficiency improvement may sometimes be a concern for decision makers, but well-defined objectives focus on achieving similarity with certain models, not positive effects on performance.

In addition, technology acquisition decisions are often made in ways resembling the "garbage can" decision model, where a purchase is followed by analysis of possible uses (Cohen, March, Olsen 1972), and buyers need to look for useful applications only after they have acquired the technology. Although the approach may look suspicious, countering the assumed rationality of decision making processes, it is wide-spread due to the complexity of ICT solutions and "sales methodologies" used by technology vendors (Bosworth 1995). In such cases, decision about the selection of means seems to precede the choice of ends – but a careful praxiological analysis would reveal the obvious existence of other ends, which guided the selection, although sometimes they could be classified as unethical… Apart from this, even technologies acquired to satisfy well-specified requirements, could in the course of implementation and everyday use be applied to other problems, thanks to their previously described, socially constructed nature.

Finally, there are cases where ethical motives justify the technology selection: similarly to the case of enforced compliance, certain decisions are made in order to avoid possible negative consequences. It is difficult (if possible at all) to evaluate effectiveness of investments in high-tech tsunami alert system, terror prevention measures at airports or environment-friendly manufacturing – economic calculations may speak against the expenditures. In the domain of workplace ethics, employers decide to adopt technologies, which make their companies more secure places to work, for example by purchasing low radiation computer displays. Even though such decisions could be interpreted as potentially increasing worker satisfaction, retention rates and long-term productivity, the economic arguments fail to capture important dimensions of managerial decisions. Value of human life escapes productivity calculus, and ethical postulates limit search for economic rents. Analysis of ICT effectiveness needs therefore to acknowledge the multiplicity of ends, extending beyond the originally assumed financial motives.

9. Discussion

The socially constructed view of technology emphasizes the ambiguity of new technologies, which may have diverse applications and changing characteristics. The overview of possible motivations, guiding ICT purchases, demonstrates in turn limitations of the assumptions of financial community and quantitatively-oriented researchers. The Misesian perspective on decision making processes points to the variety of possible ends, stating that "an end is everythink which men aim at" (Mises 1996: 93). Its heritage is clearly visible in system sciences, sociology or strategic management, and should not be forgotten when discussing technologies. Effectiveness is related to specific means-end relationships, measuring how certain means help satisfy given ends, not prescribing or evaluating the ends, which motivate human action.

Critical analysis of ICT effectiveness discourse reveals also subjective aspects of the means-end relationships. They depend not only on value systems (manifested in accepted ends), but also on meanings, interpretations of physical objects, ideas about possible applications, and their embeddedness in networks of social relations. Similarly to the postmodern interpretation of technology as text, new technologies are socially constructed by users, but still preserve their original tangible characteristics. This ambiguity poses a challenge for traditional analytical frameworks of praxiology, which used to deal with static, precisely defined means – but while postmodern scholars may fallibly suggest the demise or irrelevance of praxiological perspective, the reflection presents actually an interesting point of departure for further appropriation of the frameworks, not their rejection.

Literature

Abrahamson, E. 1991. Managerial Fads and Fashions: The Diffusion and Rejection of Innovations. Academy of Management Review, 16, 3, pp. 586–612

Anandarajan, A., Wen, H.J. 1999. Evaluation of Information Technology Investment. Management Decisions, 37, 4, pp. 329–337.

Bijker, W.E., Hughes, T.P., Pinch, T. (ed.). 1987. The Social Construction of Technological Systems. New Directions in the Sociology and History of Technology. MIT Press, Cambridge.

Bosworth, M.T. 1995. Solution Selling. Creating Buyers in Difficult Selling Markets. McGraw–Hill, New York.

Carr, N.G. 2003. IT Doesn't Matter. Harvard Business Review, May 2003, pp. 41–49.

Cohen, M.D., March, J.G., Olsen, J.P. 1972. A Garbage Can Model of Organizational Choice. Administrative Science Quarterly, 17, pp. 1–25.

Coombs, R., Knights, D., Willmott, H.C. 1992. Culture, Control and Competition: Towards a Conceptual Framework for the Study of Information Technology in Organizations, Organization Studies, 13, 1, pp. 51–72.

Daft, R.L. 1978. A Dual-Core Model of Organizational Innovation. Academy of Management Journal, 21, 2, pp. 193–210.

DiMaggio, P.J., Powell, W. 1983. The Iron Cage Revisited: Institutional Isomorphism and Collective Rationality in Organizational Fields. American Sociological Review, 48, pp. 147–160.

Faust, M., Bahnmüller, R. 1996. Der Computer als rationalisierter Mythos – vom Nutzen institutioneller Organisationstheorie für die Analyse industrieller Rationalisierung (The computer as a rationalized myth – about benefits of institutional organization theory for the analysis of industrial rationalization). Soziale Welt, 2, pp. 129–148.

Feldman, M.S., March, J. 1981. Information in Organizations as Signal and Symbol. Administrative Sciences Quarterly, 26, pp. 171–186.

Gasparski, W. 1991. Understanding Design (in Polish). Institute of Industrial Design. Warsaw.

Gordon, R.J. 2001. Technology and Economic Performance in the American Economy, NBER Working Paper, faculty-web.at.nwu.edu/economics/gordon/Tech&Econ.pdf, access date 25.06.2004.

Hatch, M.J. 2002. Teoria organizacji (Organization Theory). PWN, Warszawa.

Joerges, B., Czarniawska, B., 1998, The Question of Technology, or How Organizations Inscribe the World. Organization Studies, 19, 3, p. 363–385.

Jonas, H. 1996. Zasada odpowiedzialnoœci (The imperative of responsibility). Platan, Kraków.

Karsenti, G. 1999. La fin du paradoxe de l'informatique, Éditions d'Organisation, Paris.

Kieser, A. 1990. Bürokommunikationstechnik und organisatorische Innovation (Office communication technology and organizational innovation). Zeitschrift für Organisation, 3, pp. 171–175.

King, J.L., Gurbaxani, V., Kraemer, K.L., McFarlan, F.W., Raman, K.S., Yap, C.S. 1994. Institutional Factors in Information Technology Innovation. Information Systems Research, 5, 2, pp. 139–169.

Mises, L. von 1996. Human Action. A Treatise on Economics. Fourth Edition. Fox & Wilkes, San Francisco.

Nystrom, P.C., Starbuck, W.H. 1984. Organizational Facades. Academy of Management – Proceedings of the Annual Meeting, Boston, pp. 182–185.

Orlikowski, W. 1992. The Duality of Technology: Rethinking the Concept of Technology in Organizations. Organization Science, 3, 3, pp. 398–427.

Orlikowski, W., Hoffman, J.D. 1997. An Improvisational Model for Change Management: The Case of Groupware Technologies. MIT Sloan Management Review, 38, 2, pp. 11–21.

Orlikowski, W. 2000. Using Technology and Constituting Structures: A Practice Lens for Studying Technology in Organizations. Organization Science, 11, 4, pp. 404–428.

Picot, A., Reichwald, R., Wigand, R.T. 1998. Die grenzenlose Unternehmung. Information, Organisation und Management (Borderless Entrepreneurship. Information, Organization and Management). Gabler, Wiesbaden.

Ramiller, N.C., Swanson, E.B. 2003. Whether, When, and How to Innovate with Information Technology: What Do Empirical Studies Tell Us? Information Systems Working Paper 2–03, The Anderson School at UCLA, Los Angeles.

Rogers, E.M. 2003. Diffusion of Innovations. Free Press, New York.

Sproul, L., Kiesler, S. 1991. Connections: New Ways of Working in the Networked Organization. MIT Press, Cambridge, Massachusetts.

Triplett, J.E. 1999. The Solow Productivity Paradox: What do computers Do to Productivity. Canadian Journal of Economics, 32, 2.

Weick, K.E. 2000. Technology as Equivoque: Sense-making in New Technologies. [in:] K. Weick, Making Sense of the Organization. Blackwell, Oxford.

Application of Systemic-Structural Theory of Activity to Design and Management of Work Systems

Gregory Z. Bedny
Essex County College, Newark, NJ, USA
and
Waldemar Karwowski
University of Louisville, Kentucky, USA

1. Introduction

Praxiology and activity theories are important sources in the development of a scientific basis of ergonomics, because they provide the ultimate outcome of this interdisciplinary field, including design of equipment, HCI, efficiency of human performance, and safety. Due to the lack of a viable theory of ergonomics design there are no clearly developed design procedures. Even the term "ergonomics design" is used in many different ways. In the absence of a scientific basis for ergonomics design, it is performed in the realm of subjective, creative activity of the professionals. In this work, design is defined as the creation and description of ideal models of a new object for the purpose of its materialization in the future. This work suggests a new approach to ergonomics design, which is derived from the systemic-structural theory of activity. The design process is considered a sequence of interdependent stages and levels of analysis, and descriptions of activity structure; it is compared with the structure of design object. The design process includes a combination of qualitative, formalized, and quantitative stages. Applications of the systemic-structural approach of design are considered.

A famous Polish scientist, Tadeusz Kotarbinski (1965), developed praxiology as a philosophical framework. Praxiology is the study of human actions in terms of their efficiency. An action in praxiology is defined as intentional, goal directed behavior which is specific for human beings. Praxiology provides an explanation of human purposive action, and formulates principles of its efficient and normative performance (Gasparski, 1984). In order to study human actions, it develops basic concepts such as agent, action, aim, product, result, and efficiency. These notions are also important in activity theory. The latter is also an important theoretical framework in studying different forms of human praxis. Therefore, concept of action in praxiological terms has some similarity with concept of activity. Some philosophers in the former Soviet Union considered the concept of activity as being synonymous with the concept of action. For example, Shchedrovitsky wrote that Kotarbinski developed a science about activity, which he called praxiology (Shchedrovitsky, 1995). The difference between these two theoretical concepts is that one approach studies human praxis from a philosophical perspective, while the other uses a psychological perspective.

Activity theory utilizes two concepts. The first concept is activity and the second one is action. These two concepts are not sufficiently distinguished in praxiology. In praxiology, some scientists specify compound actions and simple actions (Czelakowski, 1993). A compound action is considered a sequence of simple actions that are performed in a particular order. Therefore, in activity theory, a compound action can be viewed as an activity and a simple action as an action. Praxiology and activity theory share the same object of study and utilize similar units of analysis. The interaction of praxiology and activity theory can enrich these two theoretical frameworks.

In this works we will consider a practical example of the systemic-structural activity theory application (Bedny and Karwowski, 2004a, 2004b, 2004b, 2004c; Bedny and Meister, 1997). Systemic-structural activity theory originated from a general activity theory (Leont'ev, 1977; Rubinshtein, 1957), and its applied branch (Konopkin, 1980; Kotik, 1974; Landa, 1976; Ponomarenko, 2004; Zarakovsky, 2004). At the same time, the systemic-structural activity theory is significantly different from these two approaches. This difference lies mostly in the development of more elaborated conceptual apparatus, well-developed stages and levels of analysis, and introduction of analytical methods of study.

Cognitive psychology currently faces the challenge of developing not just experimental, but also analytical methods of study. The need for this development becomes apparent when we consider the application of cognitive psychology to the problems of design in human performance, or human computer interaction and learning. The heart of the design is the analytic methods of developing models of an object when object itself does not yet exist as materialized system. It is proposed that the systemic-structural theory of activity and praxiology can be integrated to create a powerful theoretical and experimental approach that will allow us to tackle the difficult problems of human work activity and learning.

In psychology, the design is often incorrectly reduced to physical modeling of the equipment and further experimental studies of human performance. As a result, the creation of theoretical models of the object under design is typically omitted. In addition, physical modeling of equipment and additional experimentation can be used only as supplementary methods of study. These are typically used in the final stage of design.

Design is the creation of models of artificial objects in accordance with requirements and characteristics with the purpose of materializing these objects. The notion of a model is critical for design. The descriptive language employed for a model is a critical factor for design as well. The whole process of design can, therefore, be conceptualized as translating one descriptive language into another. Symbolical models are major models of the design process. The Systemic-Structural Theory of Activity (SSTA), with carefully developed units of analysis, can be considered a theoretical foundation of psychological aspects of design (Bedny and Karwowski, 2004a).

In the absence of the required taxonomy or language of description of human behavior or activity, it is impossible to develop analytic methods to study human performance. Taxonomy of human activity can be presented as structurally organized units of analysis. In cognitive psychology, where the major concept is process, it is impossible to describe the structure of activity. Therefore, cognitive psychology reduces design problems to experimental procedures. The SSTA can be efficiently used not only in experimental methods that aim to study human performance, but also in analytic methods of study. Currently, the only SSTA can be used in the design process.

The concept of psychological processes without action neglects the concept of a structure. As in physics, where light has both particle and wave characteristics, we understand cognition as both a process and as a system of actions (or other functional information processing units).

In this paper, we explore some of the basic concepts and principles of the general and systemic-structural theories of activity, describe their units of analysis, and discuss practical applications. We believe that the integration of the activity theory into cognitive psychology offers the foundation for a powerful theoretical and experimental approach to the difficult problems of designing human work.

2. Background and Framework

A. Activity Theory as a Potential Framework for Study of Work

One reason to use the activity theory is to provide a unified framework for studying people's work in ergonomics and industrial/organizational psychology. However, the degree of specification of a general theory of activity is not sufficiently adequate for this application. Therefore, in our paper, we pay attention not only to the general activity theory, but also to the SSTA. The Activity Theory (AT) originated in Russian psychology in the mid 1920s. AT argues that consciousness is developed in everyday practices. From an AT point of view, a person develops his or her consciousness by acting. Therefore, work has a major impact on the development of a person. For example, during an activity, people use different tools, including symbolic tools that produce not only individual consciousness, but also shared culture. The particular strength of AT is the simultaneous formulation of external behavior in terms of inner mental concepts and dynamics.

This unity of consciousness and behavior, embedded in a socio-historical context, is a major principle of AT (Rubinshtein, 1957; Leont'ev, 1977; Vygotsky, 1960). Primarily, Vygotsky paid attention to the socio-psychological aspects of mental development, whereas Rubinshtein and Leont'ev studied object-related aspects of mental development. For Vygotsky, the social dimension of consciousness was most important. For Rubinshtein and Leont'ev, the individual dimension of consciousness was most important. Vygotsky is considered the founder of the cultural-historical theory of development of the human mind in Russia. Rubinshtein and Leont'ev are considered authors of the object-oriented, individual activity theory where socio-psychological aspects of human activity are also considered. At the same time, these two theories cannot be considered totally independent, because they share similar features.

Semiotic mediation and the use of cultural tools, such as signs and symbols, constitute other important aspects of mental development. Under the use of cultural tools Vygotsky understands sign systems as the language, mathematical symbols, and gestures with a definite meaning that have evolved in the history of a culture. Tools emerge as a social force according to social-cultural features, not individual psychological features. Tools encode particular types of operations that implicitly impose upon the user constraints and prescriptions. In other words, tools socially determine practical actions and mental operations. Vygotsky introduced the notion of "instrumental act" to describe mental operations involved in the use of tools. Vygotsky connected consciousness with human labor and practical activity.

The cultural-historical theory of development of the human mind and activity theory share some basic ideas but they are two different theories. For example, meaning was a major unit of analysis in Vygotsky's cultural-historical theory of development of the human mind. In contrast, in AT the major units of analysis were cognitive and behavioral actions and their interactions with "actual reality" (material object). AT consists of goal-directed actions that can be both mental and practical. Mental actions include the manipulation of images and symbols, and motor actions transform real material objects. Meaning and sense are developed during interactions of object-practical and sign actions (Bedny and Karwowski, 2004b). Actions comprised from smaller units or operations are automatic and unconscious. In AT, motions are considered motor operations and mental acts are considered cognitive operations. Therefore, AT is a hierarchically organized structure. Mental and behavioral actions are major units of analysis and elements comprised from holistic activity. Each action has separate conscious goals that must be reached to attain the overall goal of the task. The task is a situation requiring the achievement of a goal in specific conditions; tasks can include problem-solving aspects. In this situation, the subject must find an unknown when they are not given the method by which to achieve the goal of the task.

The socio-cultural theory of the development of the human mind does not pay sufficient attention to motivational aspects of activity. In AT, motivation is considered a source of energy that directs actions or activity to conscious goals. The factors that determine a person's effort to reach the goal are called motives. One should distinguish between individual motive and motivation in general. Reykovski (1974) wrote that motivation includes both emotions and the directness of action to a specific goal. Western psychologists postulate that the goal has both cognitive and affective features; they consider the goal as a combination of cognitive and motivational components (Pervin, 1989). According to AT, it is a mistake not to differentiate between a motive and goal (Bedny, Meister, 1997; Bedny, Karwowski, 2004a); motive and goal create a vector, which give goal-directed character to activity.

B. Motive – Method of Performance – Goal

A goal is a cognitive (informational) component of an activity. Motivation, or motive, is an energetic component of activity. A goal should always be distinguished from the result. The result is what the subject accomplished during task performance; a goal is conscious. Motive or method of performance can be conscious or unconscious. An objectively given goal and a subjectively accepted goal do not always match each other. Even in rigorous experimental conditions, understanding the goal of an activity can vary depending on idiosyncratic factors and situational context. In self-initiated activity, the goal formation process becomes important.

C. Systemic-Structural Theory of Activity (SSTA)

Activity is organized as a self-regulative system and is continuously corrected based on feed-forward and feed-backward influences. Self-regulation has become synonymous with such notions as willpower, ego strength, volition and motivation (Kanfer, 1996; Kuhl, 1992). Self-regulation is a process not only of living beings, but of inanimate ones that lack consciousness, will, and motivation. The major challenge of self-regulating is the process of continuous revisiting of cognitive and behavioral strategies when internal or external conditions have changed. An activity is a goal-oriented, self-regulative system. Such a system forms its own goals and sub-goals, as well as dynamic criteria for activity evaluation. The concept of strategy is fundamental to self-regulation.

A description of an activity as a self-regulative system is known as the "functional analysis of activity" (Bedny et al., 2000; Bedny and Karwowski, 2004c; 2004d). When we talk about self-regulation, we describe activity functionally. In this case, function blocks constitute the major unit of analysis. Goals, motives, psychological processes, and cognitive actions are considered not in isolation, but from the perspective of the functional purposes in the structure of self-regulation. At this stage, we do not consider functional mechanisms independently, but from their inter-relationships and inter-dependencies. We consider a function block as a functional mechanism that has a specific relationship with other functional mechanisms. Any function may be realized by mental and motor actions or cognitive processes.

Function blocks are defined in an unvarying manner, but their content may, and regularly does, vary. The content of the function block can change, but the purpose of each function block in the self-regulation model is constant. The meaning of function blocks in any specific activity can be understood only in relation with other function blocks. Under these circumstances, the function block represents a coordinate system of sub-functions with specific purposes in the structure of activity. For example, some function blocks can create an image of a situation; others, a program of execution, a dynamic model of situation, or a complex function of control and correction. All function blocks are interrelated and influence each other. Bedny and colleagues developed different models of self-regulation, which are not discussed in detail in this paper (Bedny and Karwowski, 2004d; Bedny and Meister, 1997).

Another important method of study in SSTA is morphological analysis. Morphological analysis refers to the use of actions, operations (cognitive acts and motions), and members of algorithms to describe temporal-spatial and logical organization of activity. Members of algorithms are units of analysis that are used during algorithmic description of activity, and usually integrate one-four actions according to the particular purpose of an activity (these units of analysis are described later). The major purpose of this analysis is the description of the architecture of activity.

The logically organized system of mental and motor actions forms the structure of activity during task performance. Actions have a temporal dimension that begins when an individual accepts the goal of the action and is completed when the individual achieves the conscious goal of the action and evaluates the result of the action in terms of the established goal. Major units of analysis in systemic-structural theory of activity are presented in Figure 1.

Figure 1
Application of Systemic-Structural Theory of Activity to Design and Management of Work Systems

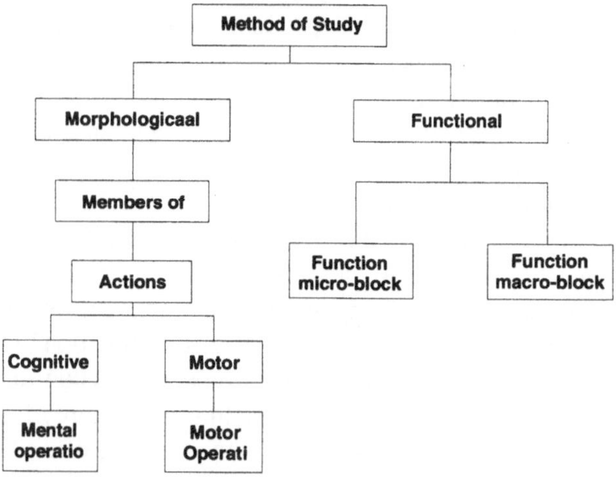

The presented material demonstrates that the major concept of cognitive psychology is the process. In contrast, major concepts of AT are actions, operations, and function blocks. Therefore, activity can be described not only as a process but also as a structure.

Actions can be classified according to different criteria. For example, according to the dominate cognitive process, actions can be sensory, simultaneous perceptual, mnemonic, and decision-making (Bedny and Karwowski, 2004c; Zarakovsky, 2004). According to the nature of the objects which are transformed by actions and method of actions performance, they can be classified in following way: object-practical (manipulation by external, physical objects), object-mental (manipulation by internal images), sign-practical (manipulation by externally presented sign), and sign-mental (manipulation by sign mentally) (Bedny, Seglin, Meister, 2000). All actions are interconnected, and their interconnection demonstrates how the principle of unity cognition and behavior in practice can be used (Bedny et al., 2001a). This suggests that cognition is not merely a process, but also a system of cognitive actions and operations. Activity during task performance is an object of study. Actions, operations, and functional blocks are all units of analysis. The object of study should be distinguished from the object of activity. The object of activity is a physical or mental product (image, mental sign) that is transformed according to the goal of the activity or action. This terminology is sometimes interpreted incorrectly during translation.

Systemic-structural analysis is possible because activity theory has developed units of analysis and corresponding methods to guide it. Actions, operations, function-blocks, and members of an algorithm of activity supply the conceptual framework for such analyses. Units of analysis are the foundation for the description of the structure of an activity and for the presentation of activity as a system.

The third method of study in systemic-structural activity theory is called the parametric method. This method concentrates on the study of different parameters of activity that are treated as relatively independent. For example, measuring separate temporal characteristics of activity is related to parametric study. In contrast, time-structure analysis of holistic activity is related to systemic analysis of activity. Another example of parametric study is cognitive task analysis, which is an attempt to describe separate cognitive processes.

The forth method is called macro-structural and micro-structural analysis. This method determines the level of analysis.

Systemic-structural analysis of activity employs distinct methods that are categorized in terms of specific criteria. Activity as an object of study embodies multiple, distinct aspects enabling the systemic-structural approach to capture the multi-dimensionality of activity during task performance. Therefore, multiple approaches must be employed to describe a single episode of activity. Consequently, we have different systemic representation of the same activity, which is called the subject of study. From this, we need to distinguish the object of study from the subject of study.

D. Stages of Systemic-Structural Analysis of Activity

Design is not a sequential process, but an iterative process with subsequent stages providing information that may cause the reconsideration of earlier stages. Systemic-structural analysis of activity entails four stages that are organized according to the loop structure principle (Figure 2). At any stage, activity can be described in different levels of detail.

Figure 2
Four stages of systemic-structural design

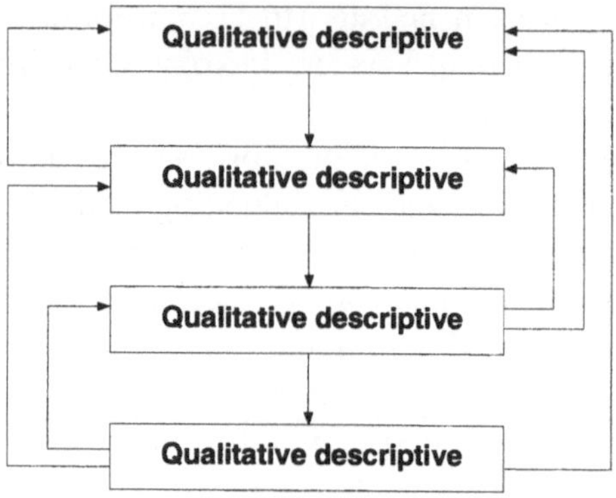

The initial qualitative stage includes a broad number of methods and usually starts with an objectively logical analysis. It may be reduced to provide a short verbal description of job performance, analysis of related production operations or tasks, and determination of their logical organization in space and time. Furthermore, a short description of the technological processes are performed including a description of all major equipment, tools, raw materials, and sequence of basic technological procedures conditions of work. Socio-cultural aspects of activity, and social norms – standards, beliefs, values, social dynamics of groups, and motivational aspects of work – are considered. The next step in the qualitative stage is the analysis of cognitive aspects of tasks. The functional analysis of activity during task performance can be performed at this stage, and experimental methods of study can also be used.

When the analyst cannot gather any more information about the task, he or she should proceed to the second analytical stage known as algorithmic analysis. An algorithm of activity is a logically structured system of actions, the order of which is determined by the set of rules defined by this algorithm. The algorithmic description of human activity divides the task into "operators" and "logical conditions." Due to the limited capacity of working memory, each member of the algorithm combines one to four actions integrated through the supervening goal. Operators consist of actions that transform objects, energy and information. Logical conditions are members of the algorithm that determine the logic of selection and realization of different members of the algorithm and include decision-making processes. Actions as units of analysis constitute the distinctive features of a human algorithm from flow charts widely used to represent human performance.

One should differentiate between prescribed and performed algorithms of activity. Prescribed algorithm is a standardized method of performance. Performed algorithm is actually used by an operator in a particular situation. It is important that these two algorithms of activity correlate as closely as possible; the more they match, the more similar and accurate the algorithmic model is to real performance.

Very often these two stages of analysis are sufficient for task analysis and design. However, in more complicated situations, or when some additional information is required, analysts can transfer to the next stages of analysis. The third stage of activity design is the description of the time structure of activity (temporal structure of activity). Time structure design is a systemic method of the description of temporal parameters of activity. This method should be distinguished from widely used methods in engineering psychology, which measure separate temporal characteristics of the operator's behavior. Operator activity cannot be presented as a number of independent reactions to independent stimuli with maximum speed. When we design the time structure of holistic activity during task performance, we need to take into consideration the pace of the performance and the opportunities to perform elements of activity simultaneously or sequentially. It is possible to build a time structure of activity only after determining with some approximation the duration of the actions involved in task performance. Before developing the time structure of an activity, it is necessary to describe the activity algorithmically. The determination of which activity elements can be performed simultaneously, and which ones should be performed sequentially, is an integral part of the time structure design.

A system of formalized rules was developed, for evaluation of elements of activity which are performed sequentially or simultaneously (Bedny, Meister, 1997). The rules for the evaluation of activity elements are based on the concentration of attention on performed actions or operations, past experiences, and existing stressful situations. These rules are important in the development of human algorithm and the time structure of activity.

The description of an activity time structure is important in the design of equipment and the evaluation of its usability. Changes in equipment configuration can probabilistically change the time structure of an activity. The specialist can evaluate and change the equipment characteristics based on time structural analysis. The time structure of activity helps the specialist evaluate the efficiency of different methods of performance (Bedny et al., 2001b). It can be also used in the evaluation of safety and training.

The last stage of systemic-structural analysis is associated with evaluation of task complexity, and other quantitative measures can be used at this stage of analysis as well. Task complexity evaluation is based on the assumption that the more complex the task, the higher the probability that it will be difficult for the performer. Therefore, between the complexity and difficulty there is a probabilistic relationship. The major problem with task complexity is developing commensurable units of measures.

Description of activity during task performance as a system of interdependent models is the purpose of systemic-structural design. These models describe the structure of the operator's activity and can be used to design equipment and develop efficient and safe methods of performance. The main idea is that changes in equipment configuration probabilistically change the structure of activity. Therefore, models of activity can be used for the evaluation of design solution.

E. Algorithmic Description of Task Performance

Algorithmic analysis of activity is related to morphological analysis. Members of an algorithm consist of actions that share a common goal Each member of the algorithm is designated by a special symbol. For example, operators are represented by the symbol "*O*" and logical conditions by the symbol "*l*."

All operators involved in the reception of information are categorized as afferent operators, and are designated with the superscripts α, as in "O^{α}". If the operator is involved in extracting information from long-term memory, the symbol μ is used as in "O^{μ}." The Operator with the symbol $O^{\alpha th}$ is involved in the performance of thinking actions with support of visual information. The symbol $O^{\mu w}$ is associated with keeping information in working memory. Operators with the symbol O^{ε} are efferent operators associated with execution, for example, the movement of a gear. In deterministic algorithms, the logical conditions designated with "*l*" have two values, zero or one. In some cases, logical conditions can be a combination of simpler ones. These simple logical conditions are connected through "and," "or," and "if-then," rules. Complex logical conditions are designated by a capital "*L*," while simple logical conditions are designated by a small "*l*. The order number of each member of the algorithm is demonstrated by a subscript number, for example, symbol $O^{\varepsilon}_{4,}$ where the number four indicates the order number of this particular member of algorithm. There is independent numbering for operators and logical conditions.

In a probabilistic algorithm, logical conditions may have two or more outputs with a probability between zero and one. As a simple example, these logical conditions may be represented in the following way. Suppose an algorithm has logical conditions with three outputs with distinct probabilities of occurrence. In such a case, the logical condition can be designated as $L_1 \uparrow^{1\ (1\text{-}3)}$ which possesses three potential values. In this case, there are three versions of output. $\uparrow^{1\ (1)}$, $\uparrow^{1\ (2)}$, and $\uparrow^{1\ (3)}$ with different probabilities. The first output has the probability 0.3, the second 0.2, and the third 0.5. Knowledge of the probability of the output may be taken into consideration in a study probability of the performance of different actions, strategies of performance, calculation of performance time of the algorithm or the components of the algorithm, and evaluation of the task complexity. Frequently, in an algorithmic description, an always-false logical condition is used, which is defined by the symbol ω. This logical condition is introduced only to make it easier to write the algorithm. It does not designate real actions performed by the subject. It always defaults to the next member of the algorithm as indicated by the arrow included in the specification of this always-false logical condition.

An arrow designates the logic of transition from one member of an algorithm to another. Thus, the algorithm exhibits all possible actions and their logical organizations and, therefore, constitutes a precise description of human performance. It describes the activity of a subject in terms of actions through which the subject attains the goal of activity. As an example, we present a fragment of an algorithmic description of a computer-based drawing task performed by Sengupta and Jeng (2003) under our supervision (Table 1, Appendix A).

Table 1
Algorithmic description of drawing task performance (fragment)

Symbol	**Member of Algorithm**	**Classification of actions**	**Time** (In millseconds)
O_1^{α}	Visual perception of the given figure	Simultaneous perceptual action	300
$l_1^{\mu} \overset{1}{\uparrow}$	Mental selection of required figure	Decision making at perceptual level. (include following operations) 1. Actualization of information from memory and, 2. Decision making	 250 370
${}^{1\,\mu}O_2$	Sustain actualized information about selected circle in working memory (abandoned option)	See description below	---
$\overset{1}{\downarrow} {}^{2\,\mu}O_2$	Sustain actualized information about selected square in working memory during performance of O_3^{ε} (correct action)	Mnemonic action of maintaining information in working memory while performing O_3^{ε} (direct connection action)	300
O_3^{ε} *	Move pointer from the starting position to the drawing toolbox	Motor positioning under visual control. (required coordination of eye and arm movement)	300
O_4^{α}	Examine area with related tools	Perceptual actions	200
$l_2 \overset{2}{\uparrow}$	Selection of required tool	Decision making at perceptual level	200
${}^{1\,\varepsilon}O_5$	Move to circle drawing icon and click (abandoned option)	See below	---
$\overset{2}{\downarrow} {}^{2\,\varepsilon}O_5$	Move to rectangle drawing icon and click (Correct action)	Motor action	380
O_6^{ε}	Move mouse to work (dotted) area, (include coordination of eye and mouse movement operations)	Complex motor actions (involve visual motor coordination Include following operations) 1. Quick eye movement to dotted area	 200

		2. Quick mouse movement to the dotted area without visual control 3. Slow mouse movement under visual control for eye and mouse position coordination	250 170
O_7^{α}	Visual evaluation of pointer position. Dynamical perceptual evaluation of mouse position during performance of O_8^{ε}	Perceptual action	780
O_8^{ε}	Motor movement of mouse pointer to the required position	Motor action	780
O_9^{ε}	Initial drawing of rectangular shape	Motor movement, which involved visual motor coordination	670
O_{10}^{ε}	Visual evaluation of shape and size of the rectangle	Perceptual action	400
$l_3 \uparrow^{3}$	Decision about correction during performance of O_{10}^{ε}		400
$\overset{1}{O}{}_{11}^{\varepsilon}$	Not to increase the size of the rectangle (abandoned action)		---
$\overset{3}{\downarrow} \overset{2}{O}{}_{11}^{\varepsilon}$	Increase size of the rectangle	Motor movement, which involved visual-motor coordination	340

The tabular form of the algorithm is carried out in the following way. On the left side of the table 1, there is a column where the above-described symbols are placed. It is a symbolic description of the algorithm or its formula. On the right side, there is a verbal description of the members of the algorithm. The symbols "*l*" or "*L*" for logical conditions in the left column include an associated arrow, numbered with a superscript, such as $\uparrow^2$. An arrow with the same number, but a reversed position must be presented in front of another member of the algorithm to which the arrow makes reference, $\downarrow^2$. Thus, the syntax of the system is based on a semantic denotation of a system of errors and superscripted numbers. An upward – pointing arrow denotes a logical state when, for example, simple logical condition, "*l*" = 1, requires skipping the following members of the algorithm until the next appearance of the downward arrow superscripted with the same number (e.g. $\uparrow^1\downarrow^1$). So the operator with the downward arrow with the same superscripted number in front of it is the next to be executed. If, for example, this is a probabilistic algorithm, one needs to skip the next appearance of the superscripted numbers for all possible arrows of the logical condition. Each superscripted number is associated with a discrete probability that needs to be represented as the transition process from the member of one algorithm to the member of another. In order to summarize this data, we present below major symbols and their description (Figure 3).

Figure 3
Explanation of major symbols consisting of a member of algorithm

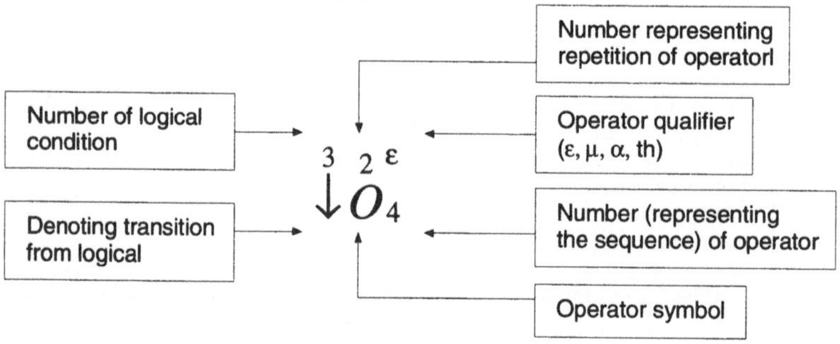

(a) Operator

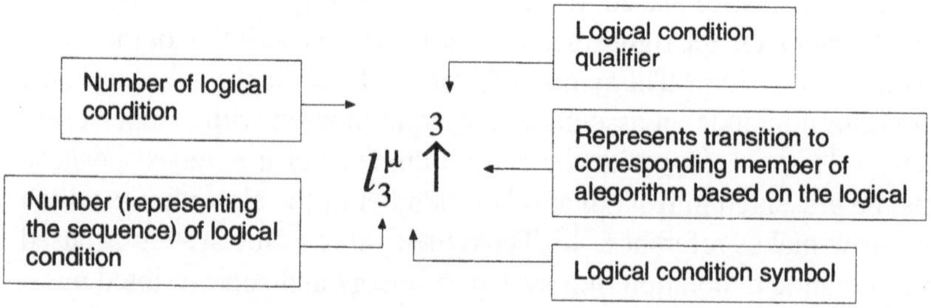

(b) Logical condition

Symbol	Denotes	Description
O	Operator	Consists of actions that transform objects energy and information
l, L	Logical condition	Determine the logic of selection and realization of different members of algorithm and include a decision making process. Can be designated as *l* or *L* *(based on combination of several logical conditions)*
á	Perceptual operator qualifier	Involvement of receipt of information for example
ì	Memory involvement qualifier	Involvement of memory actions
å	Executive operator qualifier	Involvement of executive action in terms of motor performance
th	Thinking qualifier for different members of algorithm	Involvement of thinking action

c) Description of symbols

The tabular form of an algorithm is read top to bottom. The left column with the symbolic description is called the formula of algorithm, and is presented in a vertical orientation. In some cases, a formula for the algorithm may be presented separately as a horizontal line of symbols. In this case, the formula is read from left to right. A separate formula is used when a horizontal line is presented as an abbreviated exhibition of an algorithm. In some cases, algorithms have such large versions of realizations, that experts extract only critical ones to serve as markers for the analysis. Algorithm enables an expert to describe human performance in a probabilistic manner and uncover constraints of the work process (Bedny and Karwowski, 2003a).

Following the development of the algorithm, experts then perform a psychological analysis of the algorithm, returning to a qualitative stage of analysis. Each member of an algorithm can be evaluated as a whole from qualitative and quantitative points of view.

If two members of the algorithm are performed simultaneously, they have two underlined lines (Table 1, Appendix A). The second column presents a description of each member of the algorithm (including correct and abandoned options). Usually, an algorithmic description of task presents the more often encountered abandoned options (members of algorithm). Abandoned members of algorithm, which describe unsuccessful and explorative actions, become particularly important in the description of orienting activity (Bedny et al., 2003). The third column presents classification of actions based on dominant cognitive processes. The last column presents the time of actions performed.

F. Design Time Structure of Activity during Task Performance

The rationale for the time structure analysis is that activity as a system is both a process imbedded in time and a discrete structure. Time not only reflects the duration of human performance and distinguishing features of external behavior, but it also specifies internal cognitive processes. For example, chronometrical studies play an important role in cognitive psychology (Sternberg, 1969; Zinchenko, 1978). Time structure represents a logical sequence of activity elements and duration, and the possibility of their being performed simultaneously or sequentially.

In the description of an activity, one can outline two types of analysis units. One type is called the "typical task elements," or the "technological elements," and the other is referred to as the "typical" activity element (standard). For example, "move lever," "press button," and "take reading from instrument" are all examples of typical task elements (technological units). Depending on the specificity of their performance, they can involve different activity elements. For instance, the task element "move lever" can be performed throughout different operations (motions) depending on distance, amount of effort, and precision of movement. Examples of standard elements of activity are: "detection," "decision making at sensory perceptual level," "reach," and "eye focus" from system MTM-1. In the systemic-structural theory of activity, during qualitative and algorithmic stages of analysis, specialists usually use typical elements of tasks and transform them at a later stage into typical elements of activity. Failure to distinguish between these two unit types causes the practitioner to confuse the time line chart with the time structure of activity during task performance.

For example, during a laboratory experiment, a subject was asked to fill a pin board with thirty pins. Ten of the pins had a clearly visible flute. The pins were inserted into the holes in the pin board according to the position of the flute in right and left hands. The task performance included receiving information, interpreting that information, making a decision, and performing motor actions according to specific logical sequence. Cognitive actions were combined with motor actions. At the preliminary stage, this task was described algorithmically. At the next stage, the time structure of activity during performance of this task was developed. At this stage of analysis, typical elements of task were transformed into typical elements of activity.

Figure 4a graphically displays the time structure of one member of algorithm, $O^{\varepsilon}_{4,}$ when one pin is fluted. This requires making a decision on how to turn the pin with the flute. This algorithm member is an efferent operator. The task consisted of two simultaneously performed motor actions. The motor action with the right (RH) and left (LH) hand starts at the same time. Each motor action consists of motor operations (motions). Motor operations were classified according to MTM-1 rules. Each operation is shown as a segment of the line, which together with other operations are combined into actions. The duration of the line's segment indicates the time it takes to complete elements of activity (operation or motion). These elements are building blocks to more complicated actions. Actions are integrated by high order goals and represent members of the algorithm. Therefore, activity is described as a hierarchically organized system of units. The fraction 4/9 demonstrates the probability of appearance for this particular member of an algorithm. A *dashed line* with the symbol the "MP" indicates this is a mental action. This particular action belongs to another member of the algorithm (logical conditions) and is designated by a dashed line. An action designated by a dashed line is a decision-making action at the sensory-perceptual level. This action starts at the same time as motor action. The decision-making action is associated with the decision on how to turn the left pin with the flute. The time structure of O^{ε}_{4}, when both pins are without a flute, is presented in the Figure 4b. As can be seen, the segment, which was designated by the dashed line (decision making), is eliminated in this case.

Figure 4
The time structure of one member of algorithm during instillation of pins.
a) one pin with flute;
b) both pins are without a flute.

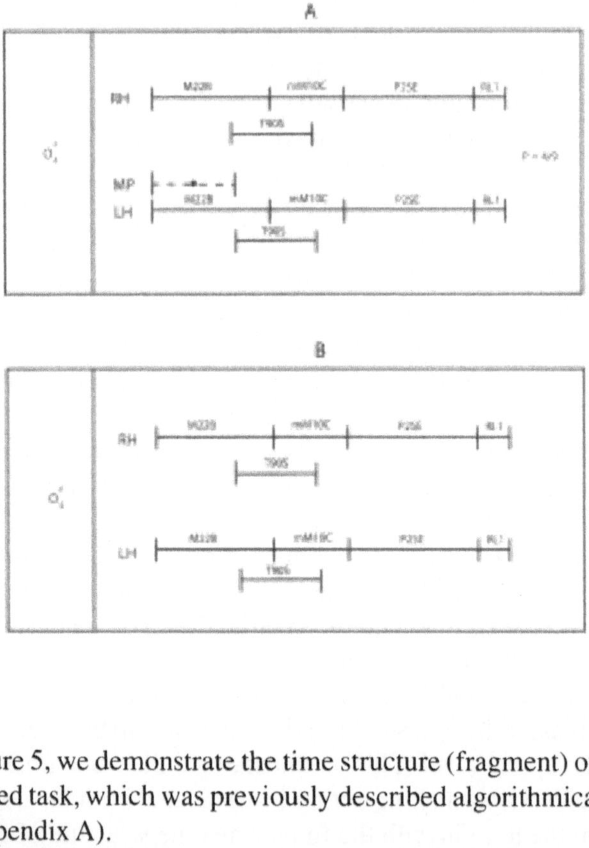

In Figure 5, we demonstrate the time structure (fragment) of the computer-based task, which was previously described algorithmically (Table 1 and Appendix A).

Knowledge about the time structure of a human-computer interaction task can provide insight to the strategy of activity during task performance. This time structure reflects efficiency of equipment, and method of performance, which can be used by an operator. During the training process, the time structure changed. Therefore, this method can also be used to create a usability evaluation and to check the efficiency of the training process. Finally, we need to pay attention to the fact that increasing complexity does not always increase time performance (compare Figure 4a with Figure 4b).

Figure 5
The time structure (fragment) of the computer-based task.

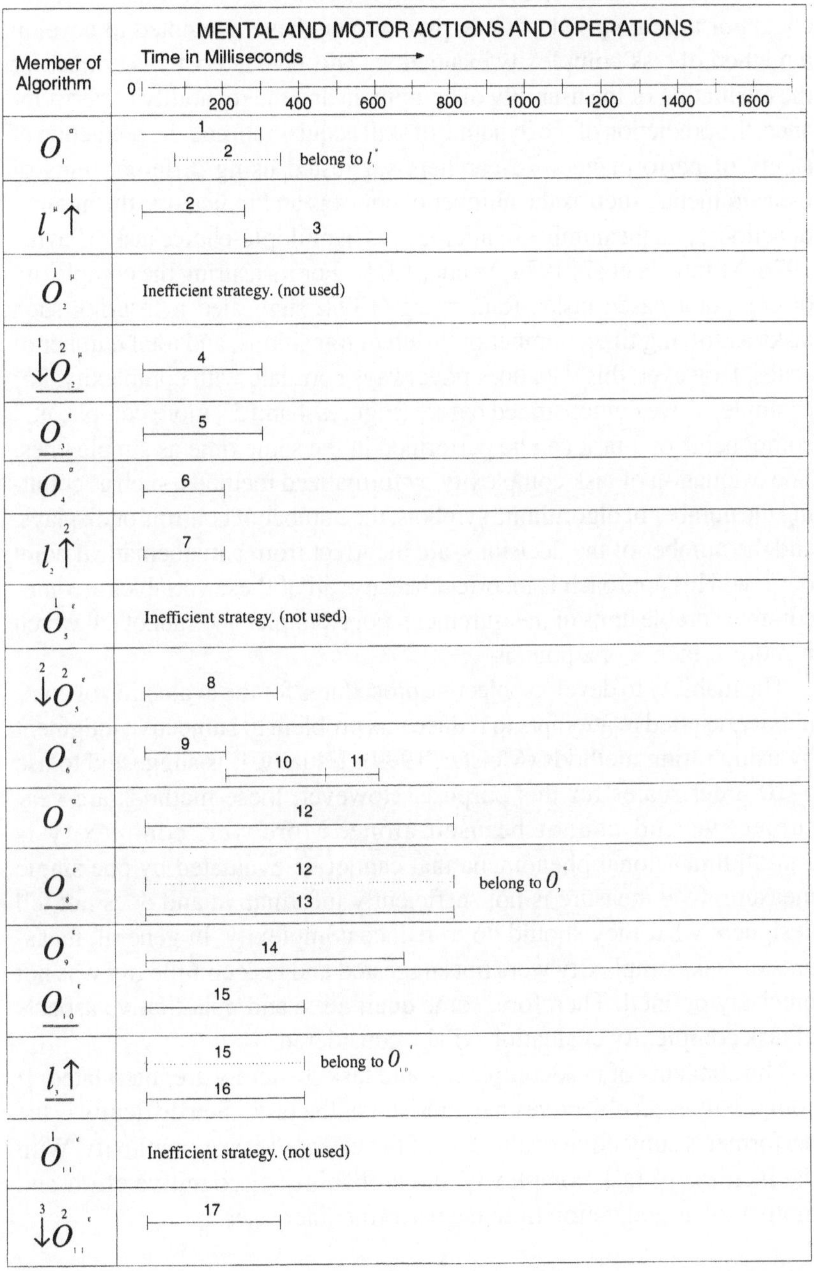

G. Evaluation of Task Complexity

Task complexity is a major factor creating a challenge for human performance. For a long period of time, practitioners have attempted to develop a method of task complexity evaluation. This characteristic is critical for the evaluation of the usability of equipment and the reliability of performance, the prediction of the dynamic of skill acquisition, and the prediction of safety of performance. Researchers suggested using different units of measurements, such as the number of controls and indicators, the number of actions, and the number of alternatives in multiple-choice tasks (Payne, 1976; Mirabella et al., 1974; Venda, 1975). For measuring the complexity of computer-based tasks, Rauterberg (1992) suggested measuring such tasks as solving time, number of different transitions, and total number of states. However, this data does not always correlate with complexity. For example, as we demonstrated before (Figures 4 and 5), more complicated components of a task can be performed in the same time as simple ones. The evaluation of task complexity by formalized methods, such as counting the number of algorithmic symbols, the number of controls or displays, and the number of the decisions, are incorrect from a mathematical point of view. This approach is incorrect because all of these variables are noncommensurable units of measurement. For example, one cannot tell which is more, 2 inches or 2 pounds.

The inability to develop objective procedures for the evaluation of complexity resulted in attempts to reduce this problem to subjective judgment by using rating methods (Meister, 1999). Usually, it is suggested to use 5–10 order scales for this purpose. However, these methods are very subjective and cannot be used alone. Moreover, complexity is a multidimensional phenomena that cannot be evaluated by one single measure. One measure is not sufficiently informative and does not tell designers what they should do to reduce complexity. In general, measures of task complexity were not suggested and task complexity was not precisely defined. Therefore, some qualitative and quantitative aspects of task complexity evaluation will be considered.

The concepts of task complexity and task difficulty are interalated. If complexity is an objective characteristic of the task, then difficulty is the performer's subjective evaluation of the effect of task complexity. With the increase of task complexity, the probability of cognitive effort and motivation mobilization from the performer increases.

Task complexity is also associated with some psycho-physiological criteria, such as the intensity of activation of the nervous system and mental efforts that require a high level of concentration. Based on the theoretical analysis of this problem, Bedny and Meister (1997) outlined five levels of activity complexity. The first, second, and third categories of complexity correspond to the first, second, and third levels of concentration of attention. For example, element "reach" can be performed with minimum concentration and is designated "RA;" the second method, designated "RB," requires an average level of concentration; the third method, designated "RC," requires a high level of concentration (Karger and Bayha, 1977). According to our study, the simplest cognitive components of activity require a third category level of complexity. For example, an operator should recognize an object and make a "yes-no" or "if-then" decision. Such components of activity should be related to the third category of complexity. Similarly, recognition of familiar objects also requires a third category level of complexity.

However, there are more complicated cognitive components of activity. For example, making decisions when the responses are unknown in advance is more complicated than if a subject were to make a decision based on two already existing alternatives. This type of decision making should be referred to as the fourth category of complexity. In the same way, cognitive actions involving attention overload or decision making in contradicted conditions, can be transferred into the fourth category of complexity. In some cases, the operator must perform in a stressful situation. Therefore, the fifth category is required. An order scale with five categories of complexity for the behavioral and cognitive components of activity was developed.

Analytical procedures of task complexity evaluation cannot be obtained based on a calculation of the number of controls, indicators, actions, or decision making and other similar criteria as was suggested before. These methods use non-commensurable units of measures. As a result, when specialists attempt to evaluate task complexity using, for example, the number of controls, they neglect the fact that sometimes, manipulation with one control can be a more complex procedure than manipulation of several controls. Similarly, the operator can perform several simple decision-making actions in one task and one complicated decision-making action in another. Hence, the evaluation of task complexity cannot be performed based on these units of measures.

It has been suggested that after developing the time structure of an activity, we can use a time interval for different components of an activity as units of complexity measurement. Using time intervals as units of measurement means that we describe qualitatively different elements of activity performed in a given time interval. If we know he time performance of a holistic task, we can mathematically calculate a means to evaluate fractions of qualitatively different components of activity during task performance. We can also classify all these fractions according to the developed qualitative criteria.

The following criteria are used to classify units of measure of task complexity: substantial characteristics of activity during a particular interval of time, complexity of these elements of activity, possibility of their performance simultaneously or sequentially, and the probability of the appearance of particular elements of activity during task performance. For example, according to the first criterion, one can distinguish the time interval devoted to mental or motor activity. The time intervals devoted to mental activity are classified based on dominant psychic processes, such as intervals of time for the perception of information, memorization, and decision-making. The complexity of the time interval depends on the level of concentration of attention during those intervals, the way the elements of activity are combined, and the existing emotional stress. As an example, we present below some formalized rules for the quantitative evaluation of the different components of activity associated with the concept of the time interval as a unit of measurement.

1. Time intervals for elements of activity requiring a lower (A), average (B), or higher (C) level of concentration of attention can be related to the first, second, and third categories of complexity (rule 1). For example, the time required for motion R30A (movement of arm 30 centimeters using method A in order to reach an object) can be related to the first category of complexity. The time that is required to reach an object using R30B and R30C should be related to the second and third categories of complexity, respectively, because R30B requires an average level of concentration and R30C requires a high level concentration.
2. The period of time when two elements of activity, which require a high level of concentration (third category of complexity), are performed simultaneously, they should be related to the fourth category of complexity (rule 2). We developed special rules, which determine the possibility to combine different elements of activity. Combination of

these kinds of elements of activity requires the highest level of resource mobilization and often increases performance time. For example, two elements of activity that require a high level of attention concentration begin at the same time. However, they are not completed simultaneously (Figures 6 elements "a" and "b"). The period of time when two elements of activity are performed simultaneously (t_1) is related to the forth category of complexity. The remaining time (t_2-t_1) is related to the third category of complexity (see figure 6 complexity interval "c").

3. The period of time when two elements of activity with different categories of complexity are performed simultaneously, should be evaluated according to the complexity of the more difficult element (see Figure 7, complexity interval "c"). In this figure element "a" has the third level of complexity and element "b" has the second level of complexity). Therefore the time interval t_1 has a third category of complexity and the remaining time interval (t_2-t_1) is related to the second category of complexity. Similarly, we develop other rules that illustrate principles of time interval evaluation.

Figure 6
Graphic interpretation of complexity according to rule 2

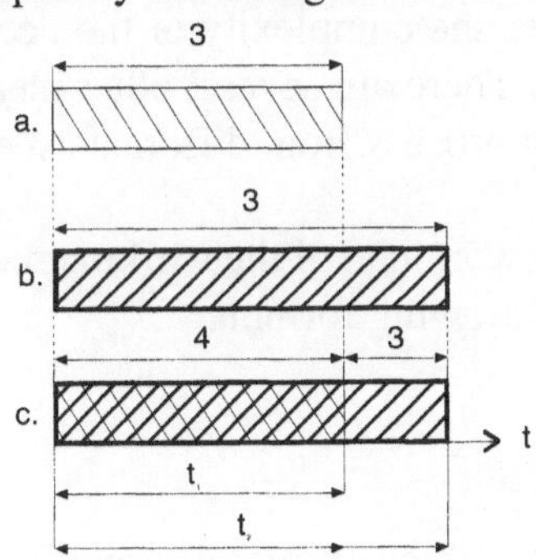

Figure 7
Graphic interpretation of complexity according to rule 3

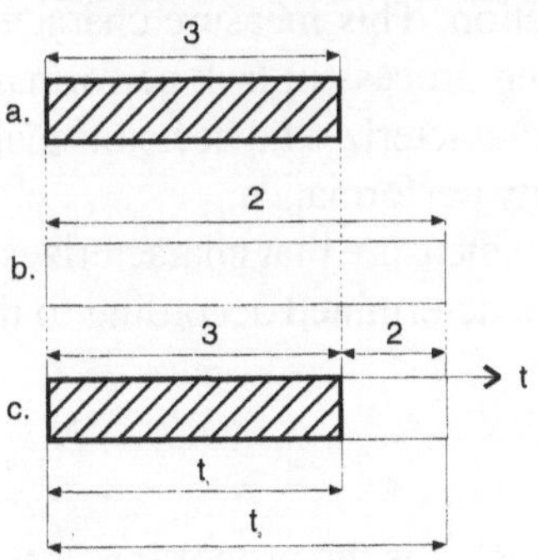

Based on these principles, it was suggested that more then twenty measurements be used. In the same way, it is possible to develop a new measurement. For example, the general duration of time for a performance of task can be calculated according to the following formula:

$$T = \sum_{j=1}^{n} P_i t_i$$

where P_i is the probability of i-th member of the algorithm and t_i is the performance time of the i-th member of the algorithm.

Time for logical conditions (decision-making processes that determine the logic of the transition from one member of an algorithm to another) can be evaluated according to the following formula:

$$L_g = \sum_{i=1}^{k} P_i^l t_i^l$$

where P_i^l is the probability of i-th logical condition and t_i^l is the performance time of the i-th logical condition. The next step is to determine, for example, the relationship between the time spent for logical conditions and the time spent for the execution of the task (algorithm execution).

$$N_\alpha = L_g / T$$

where L_g is the time for logical conditions and T is the time for algorithm execution. This measure characterizes the complexity of the decision-making process in task performance. There are several other measures that characterize the decision-making process from different aspects of activity performance.

The measure that characterizes the workload of the working memory can be determined according to the following formula:

$$N_{wm} = t_{wm} / T$$

where N_{wm} is the proportion of time in which current information is refined by the working memory and t_{wm} is the mean time for activity related to the storage of current information in working memory during task performance. T is time for algorithm execution. Table 2 summarizes the various complexity measures.

Table 2
Measures of the Complexity of Task Performance
and their psychological interpretation

Name of Measure	Formula for Calculation	Variables	Psychological Meaning
Time for algorithm execution	$T = SP_it_i$	P_i – occurrence probability, t_i – occurrence time of i-th member of algorithm	Duration of task performance
Time for performance of logical conditions	$L_g = SP_it_i$	P_i – occurrence probability, t_i – occurrence time of i-th logical conditions	Duration of decision-making process
Time for performance of afferent operators	Ta = SPata	Pa – occurrence probability, ta – occurrence time of r-th afferent operators	Duration of perceptual workload
Time for performance of efferent operators	$T_{ex}=SP_jt_j$	P_j – occurrence probability, t_j – occurrence time of j-th efferent operators	Duration of executive components of activity
Time for discrimination and recognition of distinctive features of task approaching threshold characteristics of sense receptors	$'Ta=SP_{r'}t_{r'}$	$P_{r'}$ – occurrence probability, $t_{r'}$ – occurrence time of r'-th afferent operators, characteristics of which approach threshold value (required additional EF)	Duration of perceptual process connected with weak stimuli (closely to threshold characteristics)
Proportion of time for logical conditions to time for executive activity	$N_l = L_g/T_{ex}$	L_g – time for performance of logical conditions, T_{ex} – time for executive components of activity	Relationship between decision-making process and executive components of activity

Measure of stereotypy of logical processing of information	$L_{st} = l_{st}/L_g$	l_{st} – time for stereotypy of logical processing of information, L_g – time for performance of logical conditions	Characteristics of inflexibility or rigidity of decision-making process
Measure of changeability of logical processing of information	$L_{ch} = l_{ch}/L_g$	l_{ch} – time for changeable logical processing of information	Characteristics of irregularity or flexibility of decision-making process
Measure of stereotypy of executive components of activity	$N_{st} = t_{st}/T_{ex}$	t_{st} – time for stereotypy of executive components of activity, T_{ex} – time for executive components of activity	Characteristics of inflexibility or rigidity of executive components of activity
Measure of changeability of executive components of work activity	$N_{ch} = t_{ch}/T_{ex}$	t_{ch} – time for changeable executive components of activity	Characteristics of irregularity or flexibility of executive components of activity
Scale of complexity a) algorithm b) member of algorithm	X_r - level of complexity (1,2,...5)	Based on level of concentration of attention during task performance (1 – minimum concentration, 5 – maximum)	Level of mental effort during task performance and performance of different elements. Unevenness of mental effort and critical points of task performance.
Proportion of time for repetitive logical components of work activity	$Z^l = t_{rep}/L_g$	t_{rep} – time for performance of identical logical conditions	Characteristics of habitualness of information processing
Proportion of time for repetitive afferent components of work activity	$Za = t_{rep}/Ta$	t_{rep} – time for performance of identical afferent components	Characteristics of habitualness of perceiving process
Proportion of time for repetitive efferent components of work activity	$Z^{ef} = t_{rep}/T_{ex}$ of activity	t_{rep} – time for performance of identical efferent components	Characteristics of habitualness of executive components

Proportion of time for logical components of work activity depending largely on information selected from long-term memory rather than exteroceptive information	$L_{ltm} = l_{ltm}/L_g$	l_{ltm} – time for logical components of activity whose operational nature is predominantly governed by information retrieved from the long-term memory	Level of memory workload and complexity of decision-making process
Proportion of time for retaining current information in working memory	$N_{wm} = t_{wm}/T$	t_{wm} - time for activity related to storage in working memory of current information concerning task performance	Level of workload of working memory and complexity of decision-making process
Proportion of time for discrimination and recognition of distinct features of task approaching threshold characteristics of sense receptors	Q = Ta/Ta	Ta - time for discrimination and recognition of different features of task approaching threshold characteristics of sense receptors	Characteristics of complexity, sensory, and perceptual processes
Proportion of active waiting period in total work process	$DT_w = t_w/T$	t_w - time for active waiting period in work process	Relationship between active waiting period and performance
Category of complexity of active waiting periods	X_w - 1...4	Concentration of attention during waiting period (1-minimum, 4-maximum)	Level of mental effort during active waiting period
Proportion of time for repetitive waiting periods of work activity	$Z^w = t_{wrep}/t_w$	t_{wrep} - time for repetitive waiting periods	Characteristics of habitualness of waiting periods
Measure of changeability of waiting periods in work process	$W^{ch} = t_{wch}/t_w$	t_{wch} - time for changeable waiting components in work process	Characteristics of irregularity of waiting periods
Measure of stereotypy of waiting periods in work process	$W^{st} = t_{wst}/t_w$	t_{wst} - time for stereotypy wait component in work process	Irregularity of waiting periods

We cannot discuss measures of complexity in more details in one article; these measures can be applied to different practical situations (Bedny et al., 2001b). In all cases, it was discovered that the suggested measurements can precisely evaluate the complexity of a task and relate to the difficulty of task performance.

3. Practical Example

The following example will demonstrate how we apply and develop the concept of design by using analytical procedures that are derived from the systemic-structural theory of activity. Consider the design of an underwater, unmanned vehicle. This is a sufficiently complicated and interesting example from practical and theoretical points of view.

A. Qualitative stage of analysis

The underwater vehicle must operate at a considerable depth in an area which is restricted by four underwater beacons. The operator controls the unmanned, underwater vehicle (UUV) from a control board, which is installed on a surface vessel stabilized in relation to underwater beacons or on the shore. Only one task involved in the control of the UUV movements on the seabed is considered, since this task is critical for remote control.

Remote-controlled movements of the UUV complicate the control of the vehicle position by distorting the correlation between the axis of the operator's body and that of the vehicle (see Figure 8).

Figure 8
Example of relationship between axis of operator body and axis of UUV

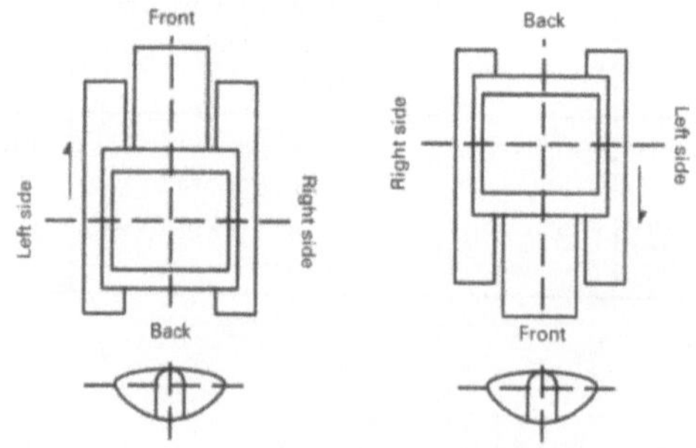

This distortion demands constant mental adjustments between these axes, which causes considerable mental fatigue. As an example, Figure 8 shows two opposite positions of the UUV in relation to the body of the operator. From this figure, one can see that the 180 degree rotation of the UUV violates the habitual relationship between the operator's body and that of the vehicle. In this situation, visual information on the screen, which installed on the horizontal panel, contradict motor manipulation by the controls. Therefore, to make the process of remote control easier, Yastrebov (1977) and Yur'nev (1975) suggested using a rotating control panel located on a surface vessel or shore to control the movements of underwater robots. Rotation of the control panel was performed by the operation using feedback from the preceding motion. However, operating such robots remains complicated. The operator handles the UUV by remote control using data from the display, which changes its position depending on the position of the control panel. Hence, axes of the UUV that are presented on the display are independent from the axis of operator's body. For the purpose of elimination of this undesirable relationship between the axes of the UUV and the axis of the operator's body, it was suggested that a rotated control panel and an oppositely rotated display be used. Rotation of the control panel is determined not by feedback from the UUV, but by the operator's control actions. For example, when the operator rotates the UUV 90 degrees to the right, he or she simultaneously rotates, in the same direction, the control board with horizontal display. However, the display rotates at the same angle in the opposite direction. This allows one to position the operator's coordinate (body axis) position in the same way as axes of the UUV. Therefore, three design alternatives for the UUV were considered: 1) stable control panel and stable computer display; 2) rotated control panel and stable computer display; and 3) rotated control panel and oppositely rotated display. The second and third versions of equipment had an inclining and reclining panel, respectively, based on the feedback from the UUV. In the third version, rotation of the control panel was followed by rotation of the display, indicating the position of the UUV, same angle, opposite direction. The display screen has a scale mask and an icon (a marker) showing the current position of the UUV (Figure 9).

Figure 9
Plan view of control board
1 – 4 points indicating position of underwater electronic beacons; 5 – control board's horizontal panel; 6 – control board's indicator; 7 – coordinate lines; 8 – code image of UUV; 9 – left turn button; 10 – right turn button; 11 – UUV's engine switch; 12 – operator's hand rest; 13 – stick slot; 14 – stick

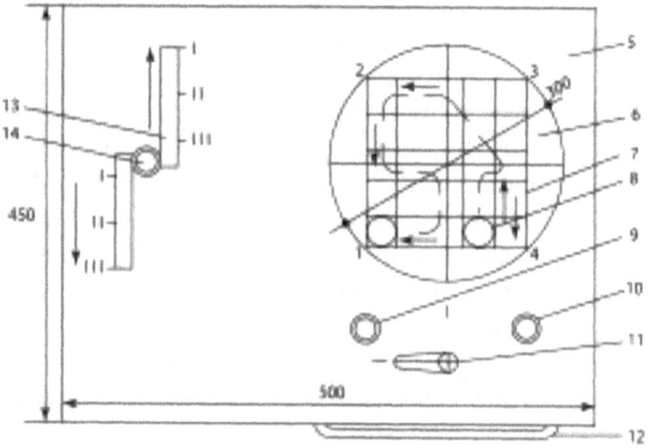

B. Algorithmic analysis

The next stage of analysis is algorithmic analysis. It starts with the algorithmic description of the task. The algorithmic description of activity during task performance (control of the UUV movements) is presented in Table 3 (Appendix A). The symbolic description of the algorithm is demonstrated on the left side of this table. On the right side of the table, there is a verbal description of the members of the algorithm, and no logical conditions or arrows, because the direction of the UUV is predetermined according to the dashed line on the screen. In Table 3, we use typical elements of a task as units of analysis (Appendix A). "Turn on the engine with the right hand" is an example of a typical element of task (technological unit of analysis).

In this task we encounter an active waiting period. An active waiting period is an interval of time when an operator may wait before starting the next element of a task after performing preliminary elements of the task. In spite of the absence of externally observed behavior, this time requires the concentration of attention. The operator is immediately ready to be involved in the performance after a waiting period.

Members of the algorithm that describe an active waiting are period designated by the symbol $O^{\alpha\omega}$.

Table 3

Algorithmic description of task performance "movement control of UUV"

Members of algorithm	**Description of members of algorithm**
O^{ε}_{1}	Turn on the engine with the right hand, moves the stick control to the moderate speed position with the left hand and return it to the hand rest position.
$O^{\alpha\omega}_{2}$	Wait until UUV moves to the first turning position.
O^{ε}_{3}	Reverse the stick control to the moderate backward speed position with the left hand and return it to the hand rest position.
$O^{\alpha\omega}_{4}$	Wait until UUV returns to the first starting position.
O^{ε}_{5}	Moves the stick control to the moderate forward speed position with the left hand and return it to the hand rest position.
$O^{\alpha\omega}_{6}$	Wait until UUV moves to the first turning position.
O^{ε}_{7}	Move the right hand to the right turn button, press and hold until UUV turns right at 90°.
$O^{\alpha\omega}_{8}$	Wait until UUV turns right at 90°.
O^{ε}_{9}	Release the right button and return hand to the rest position. Stop UUV movement by moving stick control to the neutral position.
O^{ε}_{10}	Turn on UUV movement ahead by moving stick control to the moderate speed position.
$O^{\alpha\omega}_{11}$	Wait until UUV gets to the turning point.
O^{ε}_{12}	Move the left hand to the left turn button, press and hold until UUV turns at 135°.
$O^{\alpha\omega}_{13}$	Wait until UUV turns left at 135°.
O^{ε}_{14}	Release the left button and return hand to the rest position.

$O^{\alpha\omega}_{15}$	Wait until UUV gets to the turning point.
O^{ε}_{16}	Move the left hand to the left turn button, press and hold until UUV turns at 45°.
$O^{\alpha\omega}_{17}$	Wait until UUV turns left at 45°.
O^{ε}_{18}	Release the left button and return hand to the rest position.
$O^{\alpha\omega}_{19}$	Wait until UUV gets to the turning point.
O^{ε}_{20}	Move the left hand to the left turn button, press and hold until UUV turns at 90°.
$O^{\alpha\omega}_{21}$	Wait until UUV turns left at 90°.
O^{ε}_{22}	Release the left button, press pedal "stop". Move stick control to the neutral position.
O^{ε}_{23}	Moves the stick control to the moderate forward speed position with the left hand and return it to the hand rest position.
$O^{\alpha\omega}_{24}$	Wait until UUV gets to the turning point.
O^{ε}_{25}	Press pedal "stop", moves the stick control to the moderate backward speed position with the left hand, and release pedal.
$O^{\alpha\omega}_{26}$	Wait until UUV gets to the turning point.
O^{ε}_{27}	Press pedal "stop", moves the stick control to the moderate forward speed position with the left hand, return it to the rest hand position and release pedal.
$O^{\alpha\omega}_{28}$	Wait until UUV gets to the turning point.
O^{ε}_{29}	Move the left hand to the left turn button, press and hold until UUV turns at 90°.
$O^{\alpha\omega}_{30}$	Wait until UUV turns left at 90°.
O^{ε}_{31}	Release the left button and return hand to the rest position.
$O^{\alpha\omega}_{32}$	Wait until UUV gets to the turning point.
O^{ε}_{33}	Move the right hand to the right turn button, press and hold until UUV turns at 90°.
$O^{\alpha\omega}_{34}$	Wait until UUV turns right at 90°.
O^{ε}_{35}	Release the right button and return hand to the rest position.

$O^{\alpha\omega}_{36}$	Wait until UUV gets to the turning point.
O^{ε}_{37}	Move the right hand to the right turn button, press and hold until UUV turns at 90°.
$O^{\alpha\omega}_{38}$	Wait until UUV turns right at 90°.
O^{ε}_{39}	Release the right button and return hand to the rest position.
$O^{\alpha\omega}_{40}$	Wait until UUV gets to the finishing position.
O^{ε}_{41}	Presses stop pedal by right leg, moves the stick control to the neutral position with the left hand and turn off engine with the right hand.

C. Time structure analysis.

In the third stage of analysis, the time structure of activity was developed. At this stage, only typical elements of activity were used as units of analysis. For this purpose, we applied the MTM-1 system and the action classification principles suggested by the systemic-structural theory of activity. For example, instead of using the task element (technological units) "Turn on the engine with the right hand," we used the following description "Reach engine switch 11 (distance 13 cm)." This movement is performed by method R13A, "Reach to object in fixed location." This movement is performed with minimum concentration of attention. The next motions are "Grasp the engine switch by method G1A (easily grasp)," "Move switch 11 by method A (the simplest method, distance 2.5 cm, effort less than 2.5 lb)," "Release by method RL1 (normal release by opening finger)," "Return right hand into rest position (R13E) and Grasp (G5) – contact operator hand rest 12" (Figure 9)." All described motions are considered motor operations according to the activity theory terminology. They can be combined into three simple discrete motor actions: move arm and grasp engine switch; turn the engine switch into the required position; move the arm into the rest position. This is an example of a possible standardized description of the motor elements of activity. The time structure of activity can be displayed in the form of a table or a graph. A table is used here since most of the activity elements are performed in sequence. As example the time structure for the first and the second version of UUV is presented in Table 4 (Appendix A). Time structure for these two versions of UUV are similar.

Table 4
Time structure of activity during task performance for the first and the second version of UUV

Algorithm Members	Motor Activity				Mental Activity	
	Right Hand, Right Leg	**Time (sec)**	**Left Hand, Left Leg**	**Time (sec)**	**Description**	**Time (sec)**
O^{ε}_{1}	*R13A+G1A+ M2.5A+RL1+ R13E+G5*	0.7	*R32A+G1A+ M2A2.5+ M10B2.5+RL1+ R43E+G5*	1.65	-	-
$O^{\alpha\omega}_{2}$	-	-	-	-	Active waiting	21.4
O^{ε}_{3}	-	-	*R43A+G1A+ M10B2.5+ M4A2.5+ M10B2.5+ RL1+ R24E+G5*	2.0	-	-
$O^{\alpha\omega}_{4}$	-	-	-	-	Active waiting	21.4
O^{ε}_{5}	-	-	*R24A+G1A+ M10A2.5+ M4A2.5+ M10B2.5+RL1+ R43E+G5*	1.86	-	-
$O^{\alpha\omega}_{6}$	-	-	-	-	Active waiting	21.4
O^{ε}_{7}	*R13A+G5+AP2*	0.72	-	-	-	-
$O^{\alpha\omega}_{8}$	-	-	-	-	Active waiting	10.0
O^{ε}_{9}	*RL2+R13E+G5*	-	*R43A+G1A+ M10B2.5+ M2B2.5*	1.30	-	-
O^{ε}_{10}	-	-	*M2B2.5+ M10C2.5+RL1+ R43E+G5*	1.24	-	-
$O^{\alpha\omega}_{11}$	-	-	-	-	Active waiting	20.0
O^{ε}_{12}	-	-	*R13B+G5+AP2*	0.66	-	-
$O^{\alpha\omega}_{13}$	-	-	-	-	Active waiting	15.0
O^{ε}_{14}	-	-	*RL2+R13E+G5*	0.24	-	-
$O^{\alpha\omega}_{15}$	-	-	-	-	Active waiting	39.0
O^{ε}_{16}	-	-	*R13B+G5+AP2*	0.66	-	-

$O^{\alpha\omega}_{17}$	-	-	-	-	Active waiting	5.0
O^{ε}_{18}	-	-	RL2 + R13E + G5	0.26	-	-
$O^{\alpha\omega}_{19}$	-	-	-	-	Active waiting	25
O^{ε}_{20}	-	-	R13B + G5 + AP2	0.66	-	-
$O^{\alpha\omega}_{21}$	-	-	-	-	Active waiting	10
O^{ε}_{22}	LM	-	RL2 + R13A + G1A +M10B2.5 M2C2.5	1.01		-
O^{ε}_{23}	-	-	M2B2.5 + M10C2.5 + R43E + G5	1.17	-	-
$O^{\alpha\omega}_{24}$	-	-	-	-	Active waiting	27.0
O^{ε}_{25}	LM LM	0.5	R43A + G1A + M10B2.5 + M4B2.5 + M10C2.5	1.77	-	-
$O^{\alpha\omega}_{26}$	-	-	-	-	Active waiting	27.0
O^{ε}_{27}	LM	0.25	M10B2.5 + M4B2.5 +M10C2,5 + RL1+ R43E + G5	1.63	-	-
$O^{\alpha\omega}_{28}$	-	-	-	-	Active waiting	27.0
O^{ε}_{29}	-	-	R13C + G5 + AP2	0.72	-	-
$O^{\alpha\omega}_{30}$	-	-	-	-	Active waiting	10.0
O^{ε}_{31}	-	-	RL2 + R13E + G5	0.26	-	-
$O^{\alpha\omega}_{32}$	-	-	-	-	Active waiting	20.0
O^{ε}_{33}	R13C + G5 + AP2	0.72	-	-	-	-
$O^{\alpha\omega}_{34}$	-	-	-	-	Active waiting	10.0
O^{ε}_{35}	RL2 + R13E + G5	0.26	-	-	-	-
$O^{\alpha\omega}_{36}$	-	-	-	-	Active waiting	20.0
O^{ε}_{37}	R13A + G5 + AP2	0.72	-	-	-	-
$O^{\alpha\omega}_{38}$	-	-	-	-	Active waiting	10.0
O^{ε}_{39}	RL2 + R13E + G5	0.26	-	-	-	-
$O^{\alpha\omega}_{40}$	-	-	-	-	Active waiting	20.0
O^{ε}_{41}	LM LM R13A + G1A + M2.5A + RL1 + R13E + G5	0.7	M10A2.5 + M2B2.5 + RL1 + R32E + G5	1.1	-	-

D. Quantitative evaluation of task complexity

The analysis of the time structure of activity did not show preferences for the first or second versions of the UUV, because the time structure for both versions is the same. Hence, at the fourth stage, task complexity was evaluated for all three versions of equipment. At this point, all qualitative and temporal data are transformed into quantitative measures. Not all measures of complexity showed differences in complexity. In Table 5 (Appendix A), measures of complexity are presented. Let us consider only those measures which demonstrate differences in task complexity.

Table 5
Quantitative Measures of Evaluating Task Complexity
for Manipulation of UUV on Underwater Surface

Name of Measure	1st Version	2nd Version	3rd Version
Time for algorithm execution (seconds)	311.8	311.8	309.2
Time for performance of logical conditions	0	0	0
Time for performance of afferent operators	0	0	0
Time for performance of efferent operators (T^{ex})	16.8	16.8	14.2
Time for discrimination and recognition of distinctive features of task at threshold level	0	0	0
Proportion of time for logical conditions to time for executing activity	0	0	0
Measure of stereotype of logical processing information	0	0	0
Measure of changeability of logical processing of information	0	0	0

Measure of stereotypy of executive components of activity	1	1	1
Measure of changeability of executive components of work activity	0	0	0
Scale of complexity (except active waiting period) a) algorithm b) member of algorithm Oe_9; $O\ e_{10}$; $O\ e_{12}$; $O\ e_{14}$; $O\ e_{16}$; $O\ e_{18}$; $O\ e_{20}$; $O\ e_{22}$; $O\ e_{23}$; $O\ e_{25}$; $O\ e_{27}$; $O\ e_{29}$; $O\ e_{31}$; $O\ e_{33}$; $O\ e_{35}$; $O\ e_{37}$; $O\ e_{39}$; $O\ e_{41}$;	2 2;2;2;1;2; 1;2 2;2;2;2;2 1;2;1;2;1; 1	2 2;2;2;1;2;1 2;2;2;2;2;2 1;2;1;2;1;1	2 1;1;1;1;1;1 1;1;1;1;1;1 1;1;1;1;1;1
Proportion of time for repetitive afferent components of work activity	0	0	0
Proportion of time for repetitive afferent components of work activity	0	0	0
Proportion of time for repetitive efferent components of work activity	0.13	0.13	0.85
Proportion of time for logical components of work activity depending on information selected from long-term memory rather than exteroceptive information	0	0	0
Proportion of time for retaining current information in working memory	0.83	0.83	0
Proportion of time for discrimination and recognition of distinct features of task approaching threshold characteristics of sense receptors	0	0	0

Proportion of active waiting periods in total work process	0.94	0.94	0.95
Scale of complexity of active waiting periods	1	2	2
Proportion of time for repetitive waiting periods of work activity	0	0.14	0.90
Measure of changeability of waiting periods in work process	0	0	0
Measure of stereotypy of waiting periods in work process	1	1	1

We start our analysis with quantitative measurements related to the executive components of activity. The time when an operator is actively involved in task performance for the first and second versions of the UUV is the same (16.8 sec). The third version of task time is 14.2 sec (Table 5, Appendix A). In all three versions, according to the algorithmic description, the operator performs the same actions. However, in the first and the second versions of the UUV, there are different cognitive mechanisms of regulation.

In the first version of design, all turns of the UUV, except the first turn (O^{ε}_{7}), are associated with interfering information coming from an indicator because the relationship between the operator's body axis and the UUV's axes does not coincide. This leads to a more complicated cognitive regulation of movements associated with the displacement of the hand toward the left or the right button. For example, when an operator performs O^{ε}_{12}, he or she presses the left button and the UUV starts to turn at 135 degrees. However, the control board's information does not exactly demonstrate a left turn as an operator would expect to see. This becomes obvious when the operator attempts to perform a left or right turn when the UUV is moving in the opposite direction. For example, look at the area of UUV movements associated with beacon 2. When the operator presses the left button (#9), the UUV turns to the right and when he or she presses the right button (#10), it turns to the left (Figure 8). This complicates the displacement of the right or left hand movements toward right or left buttons. According to the rules previously described, movement of the hands to the left button 9 or right button 10 instead of a simpler version of movement (element RA) should be classified as more complicate movement (elements RB or even as RC). The elements RB and RC require more concentration and time for performance. In the third version, where pressing the left button (#9) always associated with left turn and pressing the right button (#10) always associated with right turn, the movements of the hand toward a right or left button always requires lower concentration and can be related to the element RA, according to the rules developed in the systemic-structural theory of activity. These specific conditions are not taken into consideration by the MTM-1 system. In reality, pressing the left or right button is always associated with the same turn of the UUV on the screen (Figure 9). The same takes place when the operator shifts the control stick for moving the UUV ahead or back, or decides to stop the UUV altogether. Manipulation with the stick in the first and second versions requires more concentration and more time for performance. It is clear that the time structure of activity in the first and second versions of the UUV is more complicated than in the third version. The existence of interfering information regarding the UUV's position means that in the first and second versions, the time to perform motor responses equals (time performance of efferent operators from O^{ε}_{9} - O^{ε}_{41} without a waiting period) 16.8 seconds; in the third versions it equals 14.2 (Table 5).

The category of complexity of algorithm or task (except the waiting period), according to a five level order scale, is equal to two for the first and second versions of the UUV (Table 5). For the third version of the UUV, this measurement equals one. The complexity of different efferent operators, according to the five-order scale, is also present in Table 5. For example, moving the left or right hand to buttons 9 and 10 in the first and second versions corresponds to the second or third categories of complexity. This can be explained because there is contradicting information coming from an indicator and the real position of the operator's body. This contradiction is eliminated in the third version. As a result, the same motor actions in the third version always correspond to the first categories of complexity. The major steps for calculation of these measures follow. The level of concentration for each motor operation (motion), the time of performance, and the proportion of time for each motor operation are determined. Based on this data, the complexity of each efferent member of algorithm is calculated.

Another measure is the "proportion of time for repetitive efferent components of work activity (Z^{ef})." In all three versions, the operator performed externally identical actions, but in the first two versions, the same actions required the operator to use totally different information (orientating components of actions). This means that external motor actions can be simpler; however, their cognitive regulations are different. Cognitive regulation of the same motor actions was changed since the UUV's axes continuously changed their position in relation to the operator's own axis. Because of this, the Z^{ef} equaled 0.13 for the first two versions, but in the third version, the operator's axis and the UUV axes coincided. As a result, the same motor actions were regulated by the same cognitive operations. Therefore, the proportion of time for the performance of identical actions increased. This significantly simplifies the control of the UUV (Table 4). If we know the total time of performance for the repetitive efferent members of algorithm and the total time for executive components of activity, in general, we can easily determine the proportion of time for repetitive efferent components. In the third version, since the axes of the UUV coincides with the axis of the operator's body, motor actions become identical not only in their external or observable manifestation, but also in their internal cognitive regulation. Non-repetitive actions only require performance of O^{ε} and O^{ε}_{41}. As a result, the proportion of time for the repetitive response components of work activity increased to 0.85. The meaning of this measure is obvious. The more diverse the performed motor actions are, the more complicated the motor component of activity.

The next important measure of complexity is N_{wm} (proportion of time for retaining current information in working memory). In the first and second design versions, an operator must remember the position of the UUV's axis toward his own. This leads to a significant load on the working memory. An operator does not need this data except when he performs O^{ε}_{14}, O^{ε}_{18}, O^{ε}_{31}, O^{ε}_{35}, and O^{ε}_{41}, returning his or her hands to the initial position. The performance of O^{ε}_{41} does not demand memory when the left hand returns. This time equals *RL* + *R32E* + *G5*, plus the right hand's time, turn off the engine. These elements of activity take 2.91 seconds. Therefore, the time when the operator retains information about the position of UUV is 16.8 – 2.91 = 13.89 seconds. From this, the proportion of time for retaining current information in the working memory for the first and second versions of the UUV is $N_{wm} = 13.89/16.8 = 0.83$. In the third design version, the operator does not need to keep the UUV's relative position in mind. Because of this, N_{wm} is zero. This is evidence of the sharply decreasing workload of the working memory for the last version. Similar calculations were made for other components of work activity (Table 5). It was demonstrated that the third design alternative is superior to the other two versions.

The waiting period of time is also an important component of task that is considered in this study. Time for the waiting period from operator's $O^{\alpha w}_{8}$ to $O^{\alpha w}_{40}$ is 295 seconds for all three versions. The total time of task performance for the first and the second versions of UUV is 311.8 seconds, and for the third version 309.2 seconds (Table 5). The proportion of active waiting period in total task performance ΔT_w is determined according to the formula:

$$\Delta T_w = t_w / T$$

Where t_w is the waiting period of time and T is the general time of task performance. This measure for all three versions is practically the same; the time for different waiting periods depends on the speed of the UUV. However, the complexity of active waiting period is different for different versions of UUV. This period of time can be evaluated in complexity according to the level of concentration of attention during the waiting period. For this purpose, an order scale was developed and special measures for evaluation of waiting period of time were introduced.

As an example, only one fragment of this analysis is presented below. In the first and second versions of the UUV, during movements of the UUV along the straight line, an operator should remember objects positioned in relation to his or her own coordinate system. This causes some additional workload for the working memory during the active waiting period. Due to this, all waiting periods (when a screen's marker was moving along the straight line at the screen) were assigned to the second category of complexity. In a preliminary study, it was discovered that a bulldozer operator's level of attention in real conditions began to rise at 2–3 meters before the turn. When the UUV is moving, the operator should predict the moment of a turn at a coordinate scale taking into account the position of the UUV toward the coordinate axis of his or her own body. Because of this, the last 2–3 meters along the straight line before the turning point demand an even higher level of concentration. Therefore, the last 3 seconds before turning correspond to the third category of complexity. The most complex turns are the waiting periods associated with $O^{\alpha w}_{28}$, $O^{\alpha w}_{32}$ and $O^{\alpha w}_{36}$. In these cases, an operator starts to predict his or her actions earlier. Therefore, the last 5 seconds, instead of 3 seconds, should be claimed as related to the third category. Based on similar analysis, it was determined that the complexity of the waiting period for the first and the second versions of the UUV was equal to the second category of complexity, and for third version was equal to the first category of complexity (Table 5).

Other interesting measures are associated with the waiting period. The first one is a measure that describes the proportion of time for repetitive waiting periods Z^w. If an operator is processing the same information during different active waiting periods, than those periods of time are considered repetitive. The proportion of time for a repetitive waiting period in the first version of the UUV is zero because different mental actions are behind externally similar waiting periods. This can be explained by the different orientation of the UUV in relation to the operator's body axis in different parts of the UUV's trajectory. As a result, different mental actions are required in order to maintain an awareness of the correct orientation of the UUV. In the second version of the UUV, the repetitive waiting period can be observed only during the turning of the UUV with a control board. According to the trajectory of movement, the operator performs five turns an angle of 90 degrees. The time required for the first turn an angle 90 degree is not repetitive, but four other periods of time for 90 degree turning are repatative. The time required to turn the UUV 90 degrees is 10 seconds. Hence, time for repatative (90 degree) turnings is 40 seconds. This is the time for the repetitive turning period. The time for the waiting period is 295 seconds for all three versions Therefore, in the second version of the UUV, Z^w was 0.14.

In the third version of the UUV, all straightforward movement connected with repetitive active waiting periods, because of the UUV and operator body excess, coincided with each other. According to the UUV trajectory, there are five 90 degree turns. In the last four turns, the operator repeats the same mental operation as in the first 90 degree turn, because the operator's body axis coincides with the UUV axes. Only the first turning of the UUV at 90 degrees, and the subsequent turnings at 135 degrees and 45 degrees, can be related to variable waiting periods of time. The duration of all three turns requires 30 seconds. Hence, the proportion of time for the repetitive waiting period Z^w in the third version of the UUV is 0.90. This measure demonstrates that cognitive activity becomes simpler during the active waiting period in the third version of the UUV than in the first and second versions, because variability of cognitive actions in this period is reduced.

When we evaluated task complexity, we eliminated logical conditions because the UUV trajectory had been predetermined. In real work conditions, trajectory is changed depending on the situation. This means that in a real situation, an operator should independently make a decision about the turning of the UUV. Therefore, logical conditions should be taken into consideration during the evaluation of task complexity. In the first two versions, decision-making related information associated with logical conditions is presented in contradictory data. This means that the task complexity for the first two versions will be even greater than in the third version if the trajectory of movement of the UUV will not be predetermined.

The comparison of the developed analytical models, and, particularly, the measures of task complexity, demonstrate the advantage of the third version of equipment. In this case, the structure of activity is simpler, the time of active manipulation is shorter, and the difficulty of task performance is reduced.

Based on the above data, the simulator for training purposes was suggested. The simulator, which corresponds to the third version of the UUV, is presented in Figure 10.

Figure 10
Simulator for underwater, unmanned vehicle

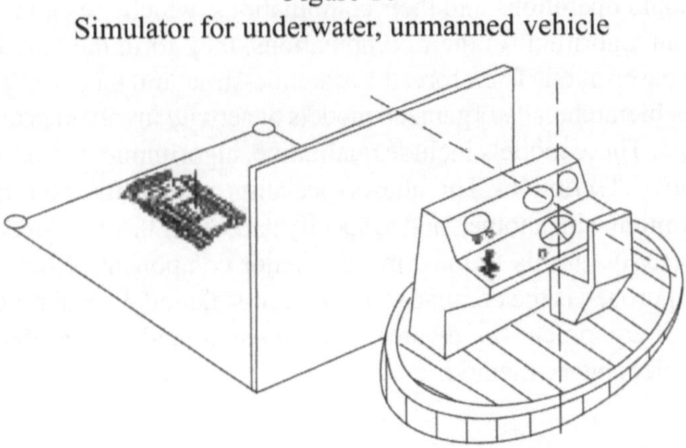

Figure 10 shows how we can organize the training process and understand design solutions for the third version of UUV. The simulator that is presented in Figure 10 consists of a rotating panel. Before the panel is a wall, and behind it is an area that models various seabed profiles. A self-propelled, proper- sized cart can move in this area. Data about the cart's position is transmitted to the opposite rotating display and the student can change the position of the cart by using corresponding controls. In this manner, the simulator helps to recreate the task involved in controlling the UUV movements along the seabed. The imitated shelf of the seabed can be changed during the training process from simple to complicated. In all three cases, it was suggested that the display be located horizontally on the panel. The panel of the board for the simulator is the same, and is presented in Figure 9.

4. Conclusion

The systemic-structural activity theory, as well as the praxiology, considers activity a system. Therefore, it is necessary to develop units of analysis to describe the elements of the system and to find out their relationship, and system structure. This allows one to describe the interrelation between the structure of activity and the configuration of material components of work. The systemic-structural theory of activity facilitates the creation of analytical methods in ergonomics design. Analogous to engineering design, the developed approach combines analytic and experimental procedures. The major units of such analysis are mental and motor actions, and operations and their combinations, which are called members of an algorithm. In their combinations, they form the language of activity description. Therefore, the systemic-structural theory of activity can offer hierarchically organized models of activity involved in task performance. These models include qualitative, algorithmic, temporal, and quantitative descriptions. This allows specialists to supplement one method of description with another, and to specify and verify the developed models of activity. In this framework, the major component of the unified design approach is the comparison of theoretical models of activity with models of equipment (hardware). Experimental methods are than used as a supplementary method.

References

Bedny, G.Z. and Karwowski, W. 2004a. "Activity theory as a basis for the study of work," *Ergonomics,* V 47, no 2, pp. 134–154.

Bedny, G.Z. and Karwowski, W. 2004b. "Meaning and sense in activity theory and their role in study of human performance," *Ergonomia,* V 26, no 2, pp. 121–140.

Bedny, G.Z. and Karwowski, W. 2004c."A functional model of the human orienting activity," *Theoretical Issues in Ergonomics Science*, Special Issue: Activity Theory. Bedny Ed.,vol. vol. 5, no 4, pp. 255–274.

Bedny, G.Z. and Karwowski, W. 2004d. "The situational reflection of reality in activity theory and the concept of situation awareness in cognitive psychology." *Theoretical Issues in Ergonomics Science,* Special Issue: Activity Theory. Bedny Ed.,vol. 5, pp. 275–29.

Bedny, G.Z. and Karwowski, W. 2003. "A systemic-structural activity approach to the design of human-computer interaction tasks," *International Journal of Human-Computer Interaction*, vol. 16, no. 2, pp. 235–260.

Bedny, G.Z., Karwowski, W. and Bedny, M. 2001a. "The principle of unity cognition and behavior: Implications of activity theory for the study of human work," *International Journal of Cognitive Ergonomics*, vol. 4, pp. 401–420.

Bedny, G.Z., Karwowski, W. and Kwon, Y.-G. 2001b. "A methodology for systemic-structural analysis and design of maual-based manufacturing operations," *Human Factors and Ergonomics in Manufacturing*, vol. 11, pp. 233–253.

Bedny, G.Z. and Meister, D. 1997. *The Russian Theory of Activity: Current Applications to Design and Learning.* Mahwah, N. J.: Lawrence Erlbaum Associates, Publishers.

Bedny, G.Z., Seglin, M. and Meister, D. 2000. "Activity theory: history, research and application," *Theoretical Issues in Ergonomics Design,* vol. 2, pp. 168–206.

Czelakowski, 1993. An outline of a formal theory of action. In T. Airaksinen and W.W. Gasparski (eds.) *Practical Philosophy and Action Theory.* Transaction Publishers, pp. 171–197.

Gasparski,W. 1984. *Understanding Design: The Praxiological-Systemic Perspectives. Seaside*, CA.: Intersystems Publications.

Kanfer, 1996. R. "Self-regulation and other non-ability determinants of skill acquisition," *The Psychology of Action,* P. M. Gollwitzer and J., Bargh, Eds. New York: Industrial Press,, pp. 404–423.

Karger, D.W., Bayha, F.H. 1977. *Engineering Work Measurements*, 3rd ed., New York: Industrial Press, Inc.

Konopkin, O.A. 1980. *Psychological Mechanisms of Regulation of Activity:* Moscow: academy of Educational Science.

Kotarbinski, T. 1965. *Praxiology, An Introduction to the Science of Efficient Action.* Oxford.

Kotik, M.A. 1974. *Self-regulation and Reliability of Operator.* Tallin, Estonia: Valgus.

Kuhl, J. 1992. "A theory of self-regulation: Action versus state orient G. M. ion, self-discrimination and some applications," *Applied Psychology: An International Review*, vol. 41, pp. 97–129.

Landa, L.M. 1976. *Instructional regulation and control: Cybernetics, Algorithmizations, Heuristic in Education.* Englewood Cliffs, N.J,: Educational Technology Publications.

Leont'ev, A.N. 1997. *Activity, Consciousness and Personality*, Englewood Cliffs, N.J.: Prentice-Hall.

Meister, D. 1999. *The History of Human Factors and Ergonomics*, Hillsdale, NJ: Lawrence Erlbaum Associates, Publishers.

Mirabella, A., Wheaton, G.R. 1974. *Effects of Task Index Variations on Transfer of Training Criteria* (Rep. No NAVTRAEQUIPMENT 72-C- 0126-1) Orlando, FL: Naval Training Equipment Center.

Payne, J.W. 1976. "Task complexity and contingent processing in decision making; An information search and protocol analysis." *Organizational Behavior and Human Performance*, vol. 16, 366–387.

Pervin, L.A. 1989. "Goal concepts, themes, issues and questions," in *Goal Concepts in Personality and Social Psychology*, L. A. Pervin, Ed., Hillsdale, NJ: Lawrence Erlbaum Associates, Publishers,, pp. 473–478.

Ponomarenko, V.A. 2004. "The significance of theoretical concepts in activity theory for applied research in aviation." *Theoretical Issues in Ergonomics Science.* V5, #4, 297–312.

Rauterberg, M. 1992. "A method of a quantitative measurement of cognitive complexity," in *Human-Computer-Interaction: Task and Organization*, G. van der Veer, M. Tauber, S. Bagnara, M. Antalovits, Eds. Roma: CUD, pp. 295–307.

Reykovski, J. 1974. *Experimental Psychology of Emotions*, Ksiazka and Wiedza,.

Rubinshtein, S.L. 1957. *Existence and Consciousness*, Moscow: Academy of Pedagogical Science.

Sengupta, T. and Jeng, O.J. 2003. "Activity based analysis for a drawing task." Proceedings of the XVth Triennial Congress of the International Ergonomics Association and The 7th Joint Conference of Ergonomics Society of Korea/Japan Ergonomic Society.

Shchedrovitsky, G.P. 1995. *Selected Works*. School of Cultural Politics Publishers.

Sternberg, S. 1969. "The discovery of processing stages: Extention of Donder's method." *Acta Psychologica*, vol. 30, pp. 276–315.

Venda, V.F. 1975. *Engineering Psychology and Synthesis of Information.* Moscow: Manufacturing.

Vygotsky, L.S. 1960. *Developing Higher Order Psychic Functions*. Moscow: Academy of Pedagogical Science RSFSR,.

Yastrebov, S.Y. 1977. ed. *Underwater Works.* Russia: Ship Construction.

Yur'nev, A.P. 1975. *Unmanned underwater apparatus*. Russia: Military Publisher.

Zarakovsky, G.M. 2004. "The concept of theoretical evaluation of operator's performance derived from activity theory." *Theoretical Issues in Ergonomics Science,* Special Issue: Activity Theory. Bedny Ed.,vol. 5, pp. 313–337.

Zinchenko, V.P. 1978. "Functional structure of executive perceptual-motor actions," *in Study Cognitive and Executive actions*, V.P. Zinchenko and V.M. Munipov, Eds., vol. 16, pp. 3–40.

The Smart Questions Approach: Changing Problem Solving to Creating Solutions[1]

Gerald Nadler
and William J. Chandon
Center for Breakthrough Thinking Inc.
Los Angeles, CA
USA

1. Introduction

Problems are inescapable in life. They are great motivators for creating technologies, whether "hard," "soft," or, the most usual, a combination of both. Then the technologies that advance human and societal health, communications, and standard of living create additional problems. The search for new technologies and those developed by the curiosity and entrepreneurship of people thus continues.

Whether it's what new product to develop, what new service to offer, what to do about global warming, how to fix a failing educational system, or how to resolve an international crisis, every person, organization, and institution has a constant need to know how to solve problems and convert technologies into creative and effective solutions and systems.

Praxiology is the philosophy of doing these things – human actions needed; seeking effectiveness, efficiency, and ethicality; and techniques for solving problems and achieving goals. Current methods advocated by almost all practical philosophies of an approach to developing solutions are based on a paradigm of a "scientific" method of problem solving. Our purpose in this article is to demonstrate that current problem solving methods are themselves a, if not the, problem in developing the creative and effective solutions so desperately needed in our societies. Furthermore, we want to synthesize a significant body of research, along with long standing philosophies about human interactions, that leads to a different approach.

2. The Problems with "Scientific" Problem Solving

The many difficulties and problems faced in society today, even with the successes of the technological age we live in today, attest to the need to raise this question. Are we trying to solve our problems, to paraphrase Einstein, with the same thinking that created them?

Our research has been extensive, involving thousands of individuals and in hundreds of different circumstances, including corporate, governmental, and personal. Our research[2] indicates that the majority of the population, around 92% of people, learned and almost all organizations use "reductionism" to solve problems.

2.1 The Reductionist Approach

The reductionist (or "rational") approach derives from the Cartesian scientific thinking paradigm that relies extensively on the use of empirical evidence, logic, and reason. Problems, in the reductionist approach, are solved "scientifically," meaning through study aimed at identifying a key part or assumption, followed by collecting data about the part, then analyzing the data, proposing a hypothesis to explain what is "correct" about the part, testing the hypothesis, evaluating the results, and concluding what the "correct" knowledge about the part ought to be. The Cartesian method is based on four principles: First, everything can be divided into its component parts. Second, any of those parts can be replaced. Third, the solution of the partial problem(s) can solve the entire problem. Fourth, the whole is nothing more than the sum of its parts.

The Cartesian method uses the following approach for all types of problems – improve an accounts payable system, set up a strategic planning process, fix a manufacturing problem, develop a community plan, or create a course syllabus:

1. Something is not working right. What exactly is broken?
2. We gather data about the situation, especially the broken or missing element.

3. We analyze the data.
4. We model or chart the data in order for others to understand it also.
5. We attempt to determine logical conclusions about precisely what's wrong or what lies at the root cause of the problem, based on the data.
6. We examine technologies and solutions to previous similar problems trying to be creative and develop a solution for correcting the root cause of the problem.
7. We implement the solution(s) that best fits this problem.
8. We apply this solution efficiently and quickly.
9. We move on to the next problem.

Almost all the professional literature on problem solving, planning, design, and related fields, the concerns of praxiology, state that this logic is *the* way to proceed.

2.2 Why Reinvent the Wheel?

A corollary fallacy is the notion that many problems or problem elements are identical or at least similar. This leads to the belief that problems can often be resolved by transferring, adapting, or grafting the technology or solution used in one problem directly onto the solution of another problem. For simple problems, such as a leaking faucet, a car noise, or a lack of letterhead stationery in our office, a mass-produced solution is fine. However, most of us need to address problems and issues that are far more complex, where the decision-making and desired implementation are far more difficult for an individual or group.

For example, a hospital wanted to improve its massive medical record-keeping system because it was overloaded, slow to respond to requests, and inefficient. After spending a lot of time and money using Reductionist thinking, the hospital heard about another hospital that had adopted a high-tech and award-winning solution to their own medical record keeping problems. The latter hospital's solution involved extensive use of computers and software. The CEO of the hospital in question asked the consultant in the first hospital to adopt the same hi-tech solution for their medical record keeping problem – he thought "there was no need to reinvent the wheel."

However, while the two hospitals were similar in size, medical services offered, socio-economic communities served, and financial condition, a great deal of time and effort might be wasted, and resistance could be engendered if an attempt was made to graft the same technological solutions to this second one. In fact, their medical records problem reflected much larger issues that had to do with several other processes within the hospital. Their problem and the needed solution were not simply a matter of installing the technology-based system of the previous hospital.

Clearly, the Cartesian method of thinking has contributed much to the world. Descartes' emphasis on analysis and empirical study led to the truly significant advances in the fields of medicine, architecture, engineering, astronomy, and life sciences that have brought us to our modern era.

2.3 The Problem with Reductionist Problem Solving

The reductionist thinking process was not intended to be the be-all and end-all of human thought. This scientific approach seeks to simplify the problems, to eliminate variables and complexities by finding the root causes and trying to patch over them. But business, societal, technological, and organizational problems have tons of variables and are dynamic and constantly in flux. Some of the major flaws with reductionist thinking for significant (non-routine) problems include:

1. *Unrelated problems cannot be treated as being similar.* Contrary to reductionist thinking no two situations can be the same. The people involved are always different; the organizations and culture are always different; and the circumstances are always different. Given this, reductionist thinking will invariably fall short in implementing the right solution for a wide range of problems.

2. *Subdividing problems into their parts does not create effective solutions.* Problems are complex, with tight interdependencies among the various parts of the system. Solving problems by replacing just a few parts also often leads to solution failure and other "unanticipated" problems. It is correct that "the sum of the parts is greater than the whole."

3. *Data collection and analysis about "the" problem is always incomplete and far too often about the wrong issue.* Too many business managers and political leaders today solve problems by collecting data ad infinitum in a frantic attempt to "know everything" possible about the problem, and thus the solutions will be obvious and jump out at us. Unfortunately, most data collection translates into "analysis-paralysis." "Data lovers" become more skilled at problem analysis than they ever do in proposing and developing solutions.

There is no such thing as perfect, total information about a problem. Data is always incomplete and time-bound. *Information is just a representation of any reality*. Regardless how often people – especially managers in large companies – insist on getting "all the data," even with the most advanced statistical techniques, there will always be a gap in knowledge about the problem, let alone the solution.

Furthermore, all measurements have some dysfunction, regardless how much someone might proclaim they have accurate measures of the right things, simply because they are always just representations of a reality and there is always an element of personal, group, community, or cultural "bias" that leads to inescapable discrepancies. In other words, information is always wrong, though some may be useful.

4. Problems almost always do not have just a single solution that will work forever. In most cases, people who use reductionist thinking become so focused on solving their problem that they fixate on one solution that appears to them fitting enough to resolve the immediate issue. But time marches on, and today's seemingly appropriate solution may not work tomorrow. Believing there is just "one and only" permanent solution is simply erroneous logic. Technology changes, people change, and the circumstances of the problem and the solution change. There is no such thing as *the* answer.

5. Creativity is sought only while developing solution ideas. The type of creativity that inspires really big ideas is severely limited by all the data and critiquing that precedes the search for solution ideas. But creativity is needed in all phases of solving problems, such as determining which people to involve, assessing what is actually the "right" problem to be working on, and assuring that the intended solution is workable. In addition, almost all of the creativity literature about how to generate new ideas addresses just the one step in reductionism that says "okay, now be creative" in finding solutions to ameliorate the difficulties or causes.

6. Solutions often overly-emphasize "exciting" new technologies. "When your only tool is a hammer, everything looks like a nail" is a well-known expression that applies to today's new technology. But the consequences of the "technology-trap" can be quite devastating.

For example, in the 1980s, General Motors decided they were going to automate all of their production activities. The company spent nearly $40 *billion* to install robotic technology in its facilities, only to remove almost all of it within a couple of years as the quality of their cars deteriorated.

As a result, the average GM car in the 1990s saw increased production costs in the range of $600-750 higher than its competitors to pay off this fiasco. The only instance in which putting technology first makes sense is in an R&D setting. For instance, Intel Corporation has an R&D unit that does ethnographic studies of families to discover what their problems are. The group has created a new chip product that senses movement, and which can be attached to the shoes or clothing on Alzheimer's patients who might wander or fall.

The flaws with the reductionist approach are serious, and thus Einstein was correct, *you cannot solve a problem with the same system of thought that created it.*

2.4 The Unstructured Method of Problem Solving

The universal ingredient in the unstructured approach derives from a firm belief in human creativity, that we are talented, inspired beings who have enormous energy to find new solutions to our problems. Proponents of the unstructured approach believe that problem solving can be accomplished largely through sheer willpower to come up with creative ideas.

One company we know about that highly endorsed unstructured problem solving caused the CEO to invite a group consisting of twenty-five of his best people and their families from around the world to come to one location where they would begin analyzing and developing better distribution practices for the company. He exhorted them to "be creative."

After three months of work on their own, the project was canceled after over a million dollars in costs, consulting fees (not ours), and uprooted families. Expecting cross-fertilization in an unstructured approach turned out to be fatal. It is too often nothing more than a hit or miss, trial and error game. While we endorse the creativity inherent in everyone, organizations, institutions, and individuals using an unstructured approach typically go down useless roads and create frustration, confusion, and disorganization among those involved.

3. Searching for a New Approach

We spent many years investigating and studying the work experiences of people who were considered to be among the most successful solution creators, the ones who won awards and received peer accolades for being exceptional creators of brilliant solutions. We began researching these people's activities in depth, through experiments, survey instruments, in-person interviews, and reading the literature about them. They all seemed to approach problems with a completely different mindset than reductionist thinking. The results of the research: *they intuitively employed a holistic, expansive thinking process rather than the rational, reductionist process.* Many of the leading solution creators actually told us they had to "throw away" the reductionist approach they had been taught.

Here are a number of characteristics that are consistently present among holistic problem solvers:

- Holistic thinkers consider every problem individually, uniquely, as if it were a brand new problem.
- Holistic thinkers seek to understand problems in a very broad context. They first try to expand their understanding to encompass the larger needs and purposes of solving each problem.
- Holistic thinkers use multiple mental styles of thought when considering problem situations.
- Holistic thinkers are willing to use their own intuition as well as rational analysis of the situation.
- Holistic thinkers are highly concerned with the people involved in and affected by the problem and its solution. They recognize that what would be called "soft data" is as important to creating a successful solution as collecting the so-called hard facts and data.
- Holistic thinkers are able to tolerate ambiguity as they fashion their solution. Situations are not black or white.
- Holistic thinkers integrate many ideas into their recommendations, highly customized to each specific circumstance.
- Holistic thinkers often seek out novel and creative solutions from internal and external sources that go beyond the simple changes to a situation that people expected.
- Holistic thinkers view their solution ideas in a systems context, as part of many interrelationships with other systems.

Creating Solutions, Not Solving Problems: We detected one more crucial distinction among holistic thinkers in that they reverse the entire context of problem solving: *They don't see themselves as solving problems, but rather as creating solutions*:

Problem solving is oriented toward the past. It aims to analyze what existed in the past, and pinpoint whatever is wrong with that. It seeks a single solution within a "fix-it" mentality. Once the problem is solved, the problem solving moves on to the next problem. Problem solving is fact-oriented, cold, "rational," and impersonal.

Creating solutions is oriented towards the future. It aims to understand situations in terms of where people want to be years from now before deciding what to do today. It recognizes that problems exist in time, and so solutions must be "living" solutions, i.e., adaptable, flexible, and ready to change as needs change. Solution creation is innovative as well as people-centered, warm, fluid – and effective.

Table 1 is a summary of what we have learned about the differences between reductionist and holistic problem solving.

Table 1
Comparison of Holistic and Reductionist Approaches

Holistic Solution Creation	Reductionist Problem Solving
Employs many mental models: intuitive, analytic, creative	Employs rational, empirical thought process
Future oriented; focuses on "creating solutions"	Past oriented; focuses on solving each single problem
People-centered	Fact-centered
Seeks out broad context in which to understand a problem and its potential solutions	Limits context to the problem itself
Aims to find unique, novel ideas that provide the basis for a "living solution" that can endure and change over time	Aims to find a single, immediate solution that "fixes" the problem
Recognizes that all information is "soft"	Emphasizes only "hard" data
Initially treats each problem situation as unique	Seeks similarities with other problems
Puts solutions in a system framework, recognizing interdependencies with other systems	Specifies changes only in terms of the parts of the problem

4. Introducing the Smart Questions Approach (SQA)

The *Smart Questions Approach* translates the best holistic thinkers' "gift" of intuitive precepts into a repeatable process that others could learn and use in their daily lives. SQA can help solve family conflicts, business problems of any size, large corporate strategic change, as well as local, state, and national political issues, and even large scale international conflicts.

Asking questions is actually one of the world's oldest techniques to think through an issue. Starting with the Greek philosopher Socrates, who challenged the youth to question the deepest moral and social beliefs of Athenian society, its most powerful principle is that thoughtful questions disrupt our "normal" and mostly unconscious patterns of thinking. They are the first intervention in implementing a change. That said, not all questions are "smart." Today's version of questioning is too often reductionist or based on assuming humans can be creative when simply told to do so.

SQA serves several critical purposes:

1. SQA challenges you with specific foundations and mental models to examine the mostly hidden assumptions and knowledge you have about the problem and its solution. They can cause you to move in the wrong direction, and more importantly, to miss gathering the right or "smart" information.
2. SQA inspires new idea generation and encourages you to explore and be creative about new arenas.
3. Smart questioning gets people to "open up." It boosts thought, imagination, and discussions about the future, and more willingness to pursue new directions.
4. SQA creates mindfulness or understanding that there could be more than one perspective or way of creating solutions, that innovation and creativity can be continuing parts of an organization's culture and motivation for ongoing change.

In summary, SQA presents a type, form, timing, and orientation of questions that differ from those posed in reductionism. SQA is built upon three foundation questions and a four-phase iterative process of questions that form the basis for the actions you take to explore and understand the problem, and to create living solutions. Here is an overview of the entire SQA.

The Three Foundation Questions of SQA

The three foundation questions of SQA provide a constant reminder for actions you will take during the 4 phases of SQA. The foundation questions are as follows:

SQA Foundation Question #1: How Can We Treat Every Problem Initially as Unique? No matter how similar two situations may appear on the surface, they cannot be the same:

- The people involved in every seemingly identical situation or system are *always* different in diversity, culture, and values.
- The purposes to be accomplished in every seemingly identical situation are very likely to be different, even with the same name, such as medication administration, strategic planning, public works, or inventory control.
- The technology available and appropriate to the problems' solution in seemingly identical situations is likely to differ.

This foundation question stands in sharp contrast with reductionist thinking, which produces the analogy of "If the Jones company can do it, then so can we." All analogies are never certain and this one is far too often costly and damaging to invoke. In addition, human nature is such that people tend to reject solutions imposed on them from outside, very often invoking the "not invented here" excuse. Even in so-called standardized situations – such as fast food restaurants, grocery chains, gas stations, and national consulting companies – the slight differences from one location to another is usually enough to prevent "standardized" solutions to every process.

This does not mean ideas, systems, and solutions that already exist will not be used; rather, this question develops the initial insights necessary for the specific situation before attempting to find available or new concepts may be usable, and avoids the too-frequent initial "technological imperative."

SQA Foundation Question #2: What Purposeful Information Do We Need to Create Living Solutions? The leading creators of solutions have basic several thoughts about information:

- Information is a human construct
- Information is always incomplete, inaccurate, and imprecise
- Information is not wisdom.

Information is a human construct. Information is not real; it is simply a representation of reality, a human construct in an attempt to paint a picture of reality. As such, each individual has a different sense of information. The oft-told story of how three blind men who came upon an elephant described it from the three different parts of the elephant each touched illustrates this truth.

Information is always incomplete, inaccurate, and imprecise. It is also impossible for data to be completely comprehensive and accurate. You cannot collect all the data about a problem; the issues are too complex and interdependent. Data is also time bound; data is not neutral because it is colored by those who collect and analyze it.

Information is not wisdom. Having information is sometimes confused with having wisdom. However, there are many types of information; consider the following levels:

· *Raw Data* – Raw data is collected using observations and the senses, and possibly the help of experts and various technologies. Trying to represent any reality with raw data, however, is not possible.

· *Real Information* – This level puts raw data to make sense of it, to see "the forest" and not just the "trees." Such manipulations still require many underlying assumptions about the data that were collected in the first place.

· *Knowledge* – Knowledge is complex and interactive. It has meaning. You can make decisions, take action, or decide not to take action. Knowledge adds interpretation, experience, history, and additional context to the representation of the issue.

· *Understanding* – To understand means that you know the real story told by the digested information as well as what it means and how it fits in "the big picture" of stories, morals, values, beliefs, implications, explanations, methods, and history.

· *Wisdom* – Wisdom is the ability to put your understanding to use. Wisdom is the transformation of understanding into concrete action. It applies a sense of values and beliefs—like justice and compassion—to knowledge and understanding in particular circumstances to come up with desirable results. It's about making a decision and getting into action.

This SQA foundation question points out that it is better to spend time gathering data that can help create solutions, not analyze problems. The goal is to seek purposeful information that contributes to *knowledge about and understanding* of solutions.

SQA Foundation Question #3: How can a Systems View Assure the Solution We Are Creating Will Work? Nothing exists by itself. Successful holistic problem prevention and solution creation hinge on taking into account the various interrelated elements and dimensions that comprise every situation, in other words, making sure that everything fits into a system.

A system is a mental framework for understanding how something should work and how the component parts of the system work together, including its purposes, inputs, processes, outputs, an environment, and human, physical, and information enablers (enablers refer to the things that make the system work). The benefits of a systems view of solutions are many:

- The solution has a framework for considering the recommended activities and events to occur.
- Everyone involved shares a language for discussing and describing what needs to be accomplished and how the solution will work.
- There is a shared vision that the solution will behave or operate as desired once it is installed.
- A documented structure exists for the solution after this solution is implemented and installed.
- All aspects of each element are considered, such as the values and beliefs of the organization, measures of desired performance, control of them as the solution is being operated, interfaces with other systems, and the future state of the solution.
- The odds of failure are greatly reduced.
- There is purposeful information available to help make the many decisions needed in crafting a solution.
- The basis for continued learning and change is established.

5. The Four Phases of SQA

The three SQA foundation questions are not an action plan, but rather form the mindset – background, guide posts, or continuous reminders of questioning – to the entire solution creation process. SQA is not linear; there is some artfulness and creativity, an openness to be as iterative as needed. It provides a four phase structure for asking questions, seeking information, and making decisions.

Phase #1: People Involvement – Who are the possible stakeholders and resource people to involve? No problem or solution exists without people. People are the cause of a problem or the recipients of its effects.

People's intelligence is the only way to create solutions. People are needed to implement the solution and live with the solution. The people to involve should represent the interests of all relevant stakeholders. These could be the workers, managers, leaders, family members, community members, customers, customers' customers, suppliers, shareholders, decision makers, "experts" in potential solution areas, and any outsiders who can impact the effectiveness of the solution.

Keeping the SQA foundation questions in mind gets the right people involved and ensures the future buy-in and acceptance of the solution from everyone involved. Involving people early, especially with Smart Questions, serves many valuable purposes that facilitate the change:

- Share meanings and interpretations about what needs to change
- Develop champions of change
- Foster a winning attitude among all individuals involved
- Develop capabilities for change among participants
- Build a network of change agents and resources
- Create teamwork and sharing
- Produce creative ideas
- Overcome resistance and arrogance
- Avoid past mistakes that people know they have made
- Overcome emotional, cultural, and environmental blocks
- Avoid the tendency to over-control the change effort
- Establish agreement and acceptance

Phase #2: Purposes – What are our many possible purposes to achieve and which one should be our "focus" purpose? This phase delves deeply into the "real" purposes to be accomplished. Instead of the conventional exhortation to "be sure you are working on the right problem," SQA shows how to identify the "focus" purpose. This purpose reflects the ultimate needs, desires, or intentions of the issue and its larger context.

The effective leaders we studied always developed some sort of hierarchy or ever-expanding set of purposes for each issue they encountered to open the door to many more possible solutions. Stated simply, they expand their "creative thinking space."

Phase 2 questions need to be aimed at understanding not just the specific goals behind fixing the problem, such as decrease processing time by 25%, but the larger purposes that you more likely want to achieve behind the surface goals. These larger purposes are often unstated, so this phase brings them out.

Phase #3: Future Solution – What is the "ideal" solution that will allow us to achieve our focus purpose(s) not just for now, but for the future? This phase seeks out tomorrow's solutions for achieving the focus purpose, so the Living Solution can build backwards from it. A consultant was asked by the president of a medium sized company to help the company design a factory that would double its production capacity. Before starting on such a "dream" assignment to design a state-of-the-art factory, the team found the real purpose of the project was "to develop management control systems." The ideal solution to achieve this end and its larger purposes was a new system for the current factory, rather than a brand new facility. The reductionist approach would have developed a great new factory solution – for the wrong problem!

Phase #4: Living Solution – What ideas can we install today that stay as close as possible to the Future Solution? Recognizing that the world constantly changes, there is no "one and only perfect, permanent solution" enduring forever. As a result, SQA seeks what we call a "Living Solution." It includes a timeline of expected changes to today's solution that keeps moving toward implementing the Future Solution within the time frame it should encompass, say, one, two, three, or more years. Many organizations also build in a later review, say, two years after initial installation, for developing a new Future Solution.

The Divergent – Convergent Structure of the SQA Phases

One additional perspective needs to be included: how to go about asking Smart Questions. It is based on the basic mental process described in the creativity literature. One major precept is that the universe of potential solutions is improved if you generate as many ideas as possible *for each phase*, rather than targeting just one or two that come immediately to mind. This process is known as "divergent" thinking. This is followed by "convergent" thinking to evaluate and weigh amongst them to select those that appear to offer the most promise given the situation.

SQA uses divergent/convergent thinking to intentionally expand your perception of the people to be involved, the purposes to be achieved, and the potential Future and Living Solutions. In addition, SQA adds another component the leading creators of solutions pointed out is critical – putting the many ideas generated in divergent thinking into several cohesive options or alternative options to consider as you move into convergent thinking.

The List Step – Divergence: What are many alternatives for the phase? Ask smart questions to generate as many ideas and topics as possible that are relevant to the phase you are in: People Involvement, Purposes, Future Solution, and Living Solution (PPFL). The nature and type of your questions will be slightly different from phase to phase.

The Organize Step – Organize: How can we organize the alternatives into possible options to consider? Given that the divergent thrust of the List step produces a large number of often random and unrelated ideas, this step organizes them into some pattern of options. The format for identifying the pattern will vary depending on the Phase of SQA.

The Decide Step – Convergence: Which option is most effective and creative for this phase? This step makes decisions about which option or options to select for action. A different framework might be used for evaluating each phase's options during the decide step.

6. What is a Smart Question?

Each phase of SQA gives rise to many questions as you proceed through the LOD steps. Because it is simply not possible to present all the questions to ask, here are some guidelines to ask yourself about any question you are considering asking to determine if is "smart":

- Does the question I'm asking align with the three foundation questions?
- Does the question I'm asking open and expand look-to-the-future responses and possibilities?
- Does the question I'm asking create new Smart Question-type metaphors and information sources?
- Does the question I'm asking feel like an interesting question or one that enhances perceptiveness on the part of others?
- Does the question I'm asking spark creative responses (in the sense that the question can yield many options), or other Smart Question-type questions?
- Is the question I'm asking likely to provide a way to empower individuals to use Smart Questions for creating solutions on their own?
- Is the question I'm asking likely to bring people together enthusiastically and with commitment to focus on building a desired future and getting results?

7. Summarizing SQA

The entire Smart Questions Approach of holistic thinking is diagrammed as shown in Figure 1:

Figure 1
The Smart Questions Approach for creating solutions

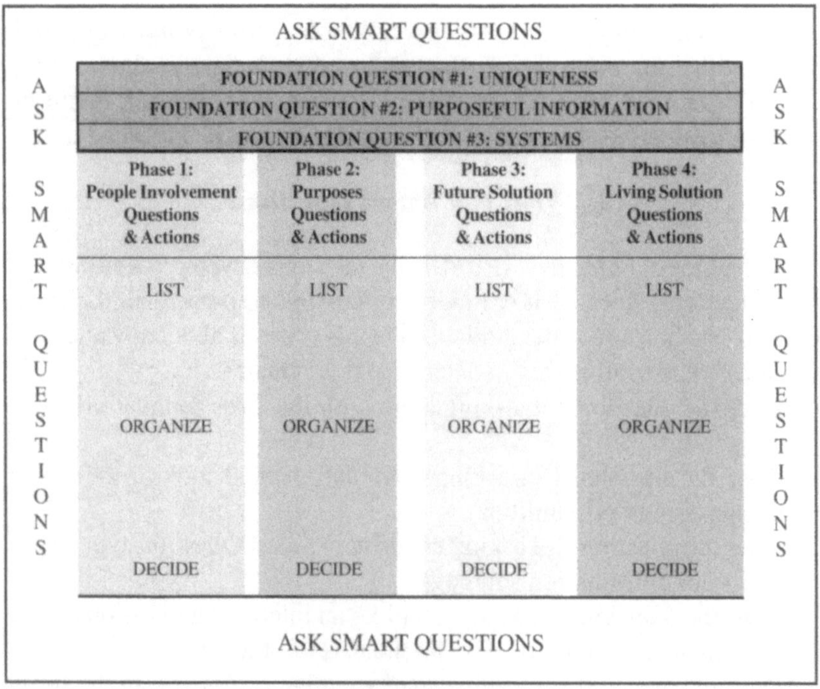

8. The Benefits of Holistic Thinking in SQA

Table 2 summarizes the leading substantive differences among the three methods of problem solving.

Table 2
Comparison of Problem Solving Approaches

Smart Questions Approach	**Reductionist Approach**	**Unstructured Approach**
Provides a semi-structured approach to problem solving that is human-centered, warm, flexible, and innovative	Provides a highly structured approach that is data-centered, cold, inflexible, and bureaucratic	Exists as a free-form approach that emphasizes creativity but lacks organization, structure, and focus.
Recognizes and respects the uniqueness of every situation as being different, independent, and deserving of its own solution	Attempts to find similarities among situations, reusing and adapting solutions that do not fit because the problems are unrelated	May or may not recognize uniqueness of a situation; or may find answers by adapting supposed creative solutions from other problems
Gets people involved extensively in the entire problem solving task, fostering ownership and investment in the outcome	Relies extensively on data and empirical evidence, often disregarding the needs and wants of the people involved	Usually involves people extensively in the solution process, but emphasizes the search for "creative" people which may not include the right people
Questions the assumptions behind the problem and the proposed solutions to ensure that the correct problem is being solved in the largest way	Accepts many assumptions as being true	May or may not examine the assumptions behind the problem or the proposed solution
Integrative approach seeks to find the largest context for understanding the problem and its solution	Divisive approach that attempts to dissect problems into small component parts; no sense of a holistic rationale behind the problem solution	May or may not succeed in recognizing a holistic solution
Accepts the possibility that there is more than one solution	Targets only one solution	May or may not develop multiple solutions or multifaceted solution
Encourages organizations to give continuous attention to living solutions that ensure long-term success	Encourages a view on fixing "the problem" and short-term success.	May or may not contain continuing focus on long-term issues and problem solving

SQA is adapted to the types of problems that organizations and institutions face today and is a disciplined and comprehensive way of getting powerful results. Its benefits include:

- It recognizes that there are always options to consider before making a decision
- It gets people involved and moves them naturally out of their current comfort zone
- It develops a willingness for paradigm shifts and delving into the unknown
- It empowers people and gives them the courage to state their convictions in a "smart" way
- It gives people a unified language for developing solutions and solving problems
- It leads to a friendly environment for change (Smart Questions Organization)
- It supports different forms of creativity during all phases of creating a living solution
- It recognizes that life is not static and there is no such thing as THE solution
- It provides everyone with a systems perception
- It makes change normal and expected because people know where it is going
- It manages team conflicts and "turf wars"
- It takes less total time using fewer resources to develop effective and creative living solutions that are very likely to be installed.

Notes

1. Adapted from Gerald Nadler and William Chandon, *Smart Questions: Learn to Ask the Right Questions for Powerful Results*, San Francisco CA: Jossey-Bass 2004.
2. For a review of much of this research, see Gerald Nadler and Shozo Hibono, *Creative Solution Finding: The Triumph of Breakthrough Thinking Over Conventional Problem Solving*, Los Angeles CA: The Center for Breakthrough Thinking Inc., 1998.

Routine Innovation and the Estrangement of the Past

Alberto Guillermo Ranea
Universidad Torcuato Di Tella
Buenos Aires
Argentina

1. Introduction

Technological inventiveness has been a highly desired child of the marriage between society and scientific progress since its first successful consummation in mid-nineteenth-century European technological settlements. In spite of angry and persistent voices urging all sorts of birth control or, even more radically, the definitive dissolution of the marriage, the progeny of this famous couple has steadily increased in number and variety in the last decades. Moreover, their children rule in present times as indisputable sovereigns, however small or irrelevant, in different kingdoms. From highly sophisticated computerized flying devices in the outer space to pencils and paper clips on a very earthly student's desk, they all share the joy of a warm welcome as if they had been long time expected by users.

The close familiarity with technological novelty has exerted undeniable influences on every aspect of human culture, most of them already discussed by specialists in different fields. Our living together with ceaselessly improved technological devices has for instance deeply affected our psychological perception of time. It is hard to remember how our lives were before, say, email, mobile phones or laptops. Like the memory of past pains, we can only have an intellectual memory of how did we experience our existences in past technological environments. They look as far removed from our present identity, as if they belonged to a foreign culture. These truistic, nay, superficial psychological remarks acquire more relevance when viewed within a different framework.

In fact, scarce attention has been paid to an analogue influence of the close partnership between humans and nonhumans on current trends in studies about science, technology and society. In them we find similar difficulties in understanding the past. It is not my intention to suggest the existence of a sort of Whiggism overspread across the studies on the history of technology, neither to propose a psychological explanation of it. However, I suspect that even for those social scientists with the subtlest historical sense it would be difficult to understand a past context when our present relationships with standing technological improvements were inexistent. More precisely, times when the benefits of technological inventiveness and frantic innovation were suspect and unwelcome among potential users.

Marks of this situation are traceable in some major contemporaneous theories on technological innovation, scientific knowledge and society. Let me arbitrarily propose three cases as exemplifying in different ways how our present context of extreme familiarity with technology might unconsciously influence even the most cleverly conceived interpretations of the past. The first is taken from the very fashionable field of socio-cultural studies of science and technology. The second focuses on the lack of homogeneity in our at first sight homogeneous attitude towards artifacts. The third comes from a more traditional approach to the relationships between metaphysics, science, and technology. I will try to suggest the existence of a subtle but effective link between all them.

2. Of Sizes and Amazements: King Hiero, Archimedes, and Us

My first example is comes from one of the most celebrated theories on techno-science during the past decade. According to it, human and nonhuman subjects, endowed with egalitarian status, constitute primeval mixed associations through processes of negotiations and alliances. These associations, or "collectives," only in a later stage appear as allegedly independent from each other under the names of "society" and "technology" – and "nature," as well. Consequently, the existence of purely human societies would be as chimerical as the existence of purely asocial artifacts.[1] I think that views like this are the natural offspring of a time when human life is intimately entangled with its technological inventions. It is hard to imagine that a similar stance could have appeared under different circumstances in the past. Our links to technological innovation are so strong and natural today, that it is easy to conclude that such mixed associations between humans and nonhumans are *per se* the very foundations of any society at any time. A theory originally aimed at an explanation of the relationships between science, technology and society in times of global markets and standing technological innovations, becomes thus an all-embracing hypothesis on the role of technical improvements in the constitution of human associations.

This scent of universality is feed with permanent references to classical Greek narratives on technical invention. When what is at issue is the philosophical or religious dimension of technology, Daedalus and Prometheus, *métis* and *tékhne* are more frequently summoned by specialists that any modern inventor's, engineer's, or technologist's approaches to the subject. It would be rather silly to deny the depth and the actuality of classical Greek reflections on the subject. However, their presence in texts on technological innovation and global diffusion of technical resources is not just an erudite event. When interspersed in researches devote to, for instance, high-tech developments, quotations from Greek or Latin sources suggest the existence of a sort of continuity between classical and present times – a continuity that gives universal validity to the local views on technology and culture held in contemporaneous researches.

Let us return to the associations between human and nonhuman "actants" (to speak the tongue of the trade). Modernity would have split them into the exclusively human realm of society on one side, and the very inert one of nature and technology, on the other. The chasm between both realms, however artificial as a construction of modern times, is quite effective. As a result, technological innovation and progress became an unstoppable fact whose rhythm and causes remain opaque to any social explanation and control. On the other side, sociologists and political scientists usually pay no attention to technicalities in discussing the role of technological issues in societies. "Socio-cultural influences on technology" and "the impact of technology on society" are different expressions of the same difficulties caused by the estrangement between both parties.

However modern this event might be, the very case on which it is supported came from Hellenistic times, when Archimedes help King Hiero in the defense of Syracuse against the Roman fleet commanded by Marcellus. Never mind: after all, "all collectives mix human and nonhuman entities together in the same way."[2]

Consequently, Archimedes' feats can be put on an equal footing than Boyle's, Hobbes's or Pasteur's. Archimedes fascinated King Hiero when he effortless drew a three-masted merchantman of the royal fleet towards him with the only help of a system of compound pulleys:

> Amazed at this, then, and comprehending the power of his art, the King persuaded Archimedes to prepare for him offensive and defensive engines to be used in every kind of siege warfare.[3]

The new relation between political power and pulleys constitute a larger collective that the one where the King had to act only politically without them. However, Archimedes decided to erase any trace of those activities because he regarded "the work of an engineer and every art that ministers to the needs of life as ignoble and vulgar."[4] As a result, the liaison between Hiero politics and Archimedean science was forever effaced. They remained in separated, independent spheres of human activities.

If we turn our attention to modern times, we find similar episodes where political collectives grew in size by joining scientific and technological resources. Boyle's and Hobbes's summoning and rejecting of the vacuum pump would represent a modern version of the Archimedean drama. Between both however the difference would be just of size. Modern technology and science, and Archimedes's geometry, differ only quantitatively. Consequently, the former are different because

> they add many more hybrids in order to recompose the social link and extend its scale. Not only the air pump but also microbes, electricity, atoms, stars, second-degree equations, automatons and robots, mills and pistons, the unconscious and neurotransmitters.[5]

There is, however, more than just a difference in size between Hellenistic and Modern collectives of politics and technologies. Let us remember the first passage by Plutarch quoted above. It reminds us Hiero's astonishment at Archimedes's inventiveness (*ekplageìs oûn ho basileùs*) as well as his meditations on technological power (*sunnoésas tês técnes tèn dúnamin*) as motives for the King's decision to ask Archimedes for assistance. To be amazed at what technology could add to political power is far from being a current feeling among today's rulers. Neither do they need a deep reflection on it in order to be convinced of how does technology add power to power. Hellenistic collectives are thus quite different from, say, post Second World War ones, and certainly much more than size is involved in their differences.

In spite of this detour through old Mediterranean landscapes, it is important to remember the importance of early modern episodes to these views on the relationships of technology with society. Moreover, "modernity" would designate

> two sets of entirely different practices... The first set of practices,..., creates mixtures between entirely new types of beings, hybrids of nature and culture. The second... creates two entirely distinct ontological zones: that of human beings on the one hand, that of nonhumans on the other.[6]

The interplay of two opposites movements of alliance and separation between politics and *tékhne* described by Plutarch would be rather a modern invention. It would explain the contradictory stance of contemporary culture towards technological progress, on the one side celebrating it as a charitable and friendly angel while, on the other, accusing it of being an alien agent of doom. This, our daily experience of simultaneous familiarity and denial of technology, is not just different in size to past experiences with artifacts and invention.

3. Space-crafts and Key Holders: Social Classes Among Artifacts

A major difference shows up when we realize that at least two different types of artifacts are involved in today's discourse on technology. On the one side, we have those projects and improvements that would lead human being – or at least, a fistful of them – to transcendental, quasi-superhuman dimensions, such as space travel and artificial intelligence. On the other side, we meet devices that, like cars and microwaves, will help us in enjoying our earthly settlements. Here, too, the generalization of one of those experiences – the former – lead to forget that the differences between both types are more than just a question of size.

The simultaneous acceptance and rejection in modern times of close links between humans and their artifacts suggests a situation akin to what René Descartes depicted over 350 years ago when speaking about mind and body. While philosophical analysis determines that there is a "real distinction" between both, we mysteriously experience their union as an impenetrable fact "en deçà de la raison".[7] The similarity is more impressive when we learn that major developments of today technology are viewed as the late offspring of Cartesian thought. Thus, artificial intelligence and thinking machines, genetic engineering and artificial life would be the main characters of a modern drama actually dawn with Descartes in mid-seventeenth-century.[8]

If the "spirit of invention" lives exclusively or, at least, most eminently, in the just mentioned high-tech developments, modern technological inventiveness's mission would be to free humanity from its miserable condition. Accordingly, technologies that contribute to a more comfortable sojourn in daily life should be despised. The religious connotations of this stance are far from imaginary:

> Technology had come to be identified with transcendence, implicated as never before in the Christian idea of redemption. The worldly means of survival were now redirected toward the other-worldly end of salvation.[9]

But millenarian redemption is not for everybody, but only for the elects. Consequently, these technologies were not meant to meet basic human needs. "They have been aimed rather at the loftier goal of transcending such mortal concerns altogether."[10] A technology for an elite, whatever the meaning of this word might have: a class or a race, a country or a corporation. No wonder the advocates of the radical estrangement between technology and society usually chose their examples among these high-tech devices.

The wells where these views on technology and human transcendent destiny have been watered are in the pages of an early twentieth-century program for future technological developments. However suggestive its title might appear, its author is far from supporting any known religious stance.[11] The World, the Flesh, and the Devil are the names for the physical, the physiological and the psychological elements of future human evolution. The colonization of space refers to the first item, while the mechanization of the body to the second. Both would contribute to a better life for human beings in a future stage:

> The dissimilarity between the conditions of life in space and on the earth would in itself be sufficient to cause perfectly normal, unassisted, evolutionary changes in human beings, but obviously spatial conditions would be more favourable to mechanized that to organic man. If he could get rid of the major part of his body [...] [t]his would give mechanized man an advantage similar to that which the

> relatively flexible and naked animal cell has over the rigidly demarcated plant. Besides, it is only in space that the potentialities of the more highly developed forms of complex minds would have an adequate field of functioning, particularly in their extended time relations.[12]

"Desire for progress," the weapon for defeating the Devil of human passions, is the essential third item in the rationally induced and controlled evolutionary future. People without it will simply stay on the Earth, while a handful of human beings will take the route marked by their rational souls, a step further into human evolution.[13]

A much less dramatic viewpoint arises when also less transcendental technologies than spacecrafts, artificial intelligence or artificial life are taken into account. Our life together with any sort of objects and devices clearly hints at a more comfortable daily stay on earth. It is irrelevant that most of them might have been originally planned for warfare. As soon as they begin circulating through our bedrooms and soccer arenas, they become natural personas of our very earthly agendas. Inventiveness in this case has no religious meaning. On the contrary, perplexity and astonishment at invention do not lead to attributing to the inventor an otherworldly status.

Moreover, some of those artifacts deserved a close attention in specialized literature on society and technology. Seat belts, hotel key holders, and automatic doors, for instance, have been the subject of rich recent analyses.[14] According to these, we delegate to them the burden of responsibility in cases where the unexpectedness of our behaviors might create difficulties for collective life. In this sense, objects like the just named are anthropomorphic, but in an unusual sense of the word: they give shape to our human actions.[15] And whatever the meaning of "human nature" might be, they too collaborate in its further evolutionary steps. In this case, however, the evolution is not aimed at transcendental goals. On the contrary, it is an earthly evolution of our species tightly linked with its nonhuman fellows. A friendly evolution embracing every consumer all round the world replaces the elitism of the other view on technology and human destiny.

When we focus our attention on this less dramatic relationship between society and technology, the above views on modernity and collectives lose part of its momentum. Let me remember the list of objects quoted above in footnote 5. The two-directions movement of modernity will firstly incorporate them into the collectives and then, in a second, opposite step, will erase their footprints in human societies. The list is quite suggestive: air pump, microbes, electricity, atoms, stars, second-degree equations, automatons and robots, mills and pistons, the unconscious and neurotransmitters. Do we share our daily lives equally with any of them? Are them all equally erasable of our human collectives as if they had no part in their constitutions?

I will not dare to answer here these questions. I would only like to express my suspicions that we have different answers if the attention is put on the exclusiveness and ceaselessness of exquisite innovation or on the different timing and goals of the use of artifacts. I do not even try to deny the existence of a continuum between both approaches. However, many important interpretations of the relationship between technology and society feed their arguments with examples from the amazement at invention and novelty. King Hiero and Archimedes, The Royal Society and the air pump, Alan Turing and machine intelligence are among the most exciting cases. But, is it also the case with, say, mills and pistons?

4. Baroque Inventors,. Universal Thinkers

These devices brings us back again to seventeenth-century Europe when, according to major today's theories on technology and society, we find the dawn of modernity in science and technology. According to a widely accepted view, the Baroque saw the achievement of the Renaissance ideal of *uomo universale* in the unity of scientist, philosopher, and technician.

> Fast bei allen grossen Forschern des barockzeitalters lässt sich in mehr oder geringerem Masse der Wille nachweisen, nicht nur die letzten metaphysischen Grundlagen ihrer Wissenschaft zu erforschen, sondern auch durch technische Apparaturen... zu illustrieren.[16]

> [Almost all great researchers of the Baroque show more or less willingness not only to examine the metaphysical principles of their discipline but also to illustrate them my means of technological apparatus.]

Descartes, Pascal, Leibniz, Galilei and Newton are the towering characters who "verkörpern personal... die Einheit von Technik, Forschung und Philosophie."[17] [embody personally... the unity of technology, research, and philosophy.]

Except for some arguable exclusion – as it is the case of Christian Huygens – there is no discussion about the importance that all these names have in the history of science and philosophy. None of them, however, measured by today's criteria, would count primarily as an inventor. Even Leibniz, in spite of the immense and undeniable importance of his writings and correspondence on technological issues – many pages of which still remain unpublished – would not be considered as an inventor with, say, strong metaphysical or logical inclinations, but quite the other way round. As the just quoted essay asserts, for Leibniz, as for the other scientists of the list, technical apparatus would have been just illustrations of their more important scientific and philosophical ideas. If it were exclusively for their contributions to technological improvements, it is reasonable to conjecture that their names would not have appeared in the Pantheon of learning.

Was there during Baroque years people who could qualify, then and now, as plain inventors? If so, have they left any written testimony about their inventing and improving inventions? Do they have recourse to metaphysical ideas for a better understanding of their activities? Do they consider their inventions as illustrations of their general conceptions on knowledge, nature or society? To satisfactorily answer all these questions would take me beyond the scope of this text. Let me mention an example: Denis Papin, a French Huguenot physician born at Blois, France on 22nd. August 1647 and whose place and date of decease are still conjectural. He is often called a mediocre scientist and inventor by historians and philosophers alike. On the basis of just a small part of the available documental evidence, he was even been described in a highly regarded essay as merely a technician to Robert Boyle.[18]

A close lecture of his long correspondence between 1692 and 1707 with Leibniz suggests however a quite different picture. Unfortunately, this is only possible still now through the manuscripts sources. There have been two editions of this important *Briefwechsel*, both at the end of nineteenth-century, but they present unforgivable deficiencies. The first edition by Ernst Gerland left unpublished almost two-third of the letters on the ground that they were devoted to just the controversy on living force.[19] The second trial was certainly more complete, but the transcriptions are unbelievable bad, the letters are arranged in a strange chronological order.[20] On top of it all, the editor attributed to Papin the well-known Leibnizian essay "Brevis demonstratio erroris memorabilis Cartesii".[21] The fortune of this edition has been very strange. Only few copies of it are still available, and major public libraries do not own it. Moreover, and surely because of these odd circumstances, Emile Ravier did not mention it in his catalogue.[22]

Papin never tried to search for the metaphysical foundations of his theories, nor has he ever regarded his inventions as merely illustrations of scientific or philosophical theories. He was eminently an inventor. For him, the construction of a machine always had priority over any theoretical consideration. Thus, when "la theorie" and "la pratique" were in conflict, he always unhesitatingly lean to the latter. It would be quite unfair to criticize him for not having found a useful connection between theory and practice in the fabrication of machines. Even the most celebrated theoretical achievements of the time, such as the "moulin à feu" of Guillaume Amontons, were never actually constructed.[23] Besides, the radical difference between theory and practice was openly accepted by the Academie Royale des Sciences – a Company that enjoyed so much prestige in the eyes of learned people like Leibniz himself. On 6th March 1704 Papin wrote to the latter:

> il y a environ trente ans que J'ay eu cette pensée et que je la proposay à Paris à Mr. Hugens, Perrault et autres membres de l'Academie des sciences: L'invention fut fort approuvé pour la Théorie; mais on ne crut pas qu'on la pût mettre en execution: et ce n'a été qu'a force de travailler à diverses choses... que les pensées me sont venues... pour surmonter toutes les difficultez de la pratique.[24]

> [... about thirty years ago I had this thought and I proposed it to Mr Hugens, Perrault and other members of the Academy of sciences. The invention was warmly approved in theory: but it was believed that it was not executable. Only by dint of working on different things... the thoughts came to me... of how to overcome all the difficulties of the practice.]

Papin has always give priority to the practice, although in several cases for different reasons. He sent to Leibniz on 22nd September 1704 his remarks on the celebrated "moulin à feu d'Amontons". He was categorical in his criticism:

> ... il y paroît un genie inventif et penetrant pour decouvrir les circonstances qu'il faut examiner afin de determiner les avantages qu'on doit attendre d'une nouvelle invention: mais je trouve aussi qu'il a peu de connoissance de la pratique. La machine de la grandeur qu'il la dêcrit est tout a fait impraticable.[25]

> [... he (Amontons) seems an inventive and penetrating genius for discovering those circumstances one need to examine in order to determine which are the expected advantages of a new invention: but I find also that he has not much knowledge of the practice. The machine, with the largeness with which he describes it, is quite impracticable.]

It is true that he enthusiastically kept a long debate with Leibniz on the measure of force between 1692 and 1700. And it is true also that he probably never understood Leibniz's criterion for the measurement of forces by their actions in vertical displacements. But his reasons did not stem from a dogmatic Cartesianism, much less from mere lack of cleverness. When he adopted a theory about the cause of gravity or on the spring of the air, for instance, it was in order to support his calculations – incidentally, frequently wrong – for evaluating the actions of concrete machines. Besides, Papin rejects Leibniz's conception of the violent effect as the right measurement of force not because of theoretical inability but on the ground of his vast and bitter experiences with high expenses and wastefulness of real machines.

But Papin was not only a projector of machines that could be seen as anticipating later highly successful devices like the steam engine. He also designed daily objects for a more comfortable life. In the just quoted letter to Leibniz of 22nd September 1704, Papin reported him about the invention of an airbed.[26] Leibniz´s answer is unfortunately lost, but from Papin's next letter (30th October 1704) we learn that Leibniz immediately order him an air cushion for his carriage.[27] On 15th January 1705, Papin announced to Leibniz:

> Je feray faire le plus tôt qu'il me sera possible le cousin pour vôtre chaize et J'espere que Vous en trouverez l'usage commode.[28]
>
> [I will make as soon as possible the cushion for your seat, and I hope you will find it comfortable.]

To Papin, success in invention meant only that the created objects do in fact work. This is equally valid for a steamboat, an air cushion or a pressure cooker. There is no purely theoretical success in a time were most people were against "tout ce qu'on appelle nouveautez"[29] [everything that people call novelties.]

Theoretical speculation becomes in his writings an *ancilla artis*. He thus situated himself on the antipodes of those who typically represent the Baroque ideal of the unity of metaphysics, science and technology. He firmly rejected as pointless Leibniz's speculations on entelechies and *vis viva et mortua* in their debate on the measure of force.[30] He focused his attention on devices that could make our painful lives on earth more comfortable. In this sense, his feats belong to the technologies of seatbelts and microwaves more than to the transcendental interpretations on innovation related with what was termed a Cartesian agenda. Moreover, his writings remind us that inventiveness and novelty have not always been celebrated as the desired children of the human condition. In our relationship with them therefore more than just a matter of size is certainly involved.

Notes

1. Latour, Bruno. *We Have Never Been Modern*, Cambridge, Mass.: Harvard University Press, 1993.
2. Latour, Bruno. *We Have Hever...*, 108.
3. Plutarch, Marcellus' Life, XIV, 7–9, trans. Bernadotte Perrin, quoted by B. Latour, *We have never...*, 109.
4. Plutarch, Marcellus' Life, XVII, 4–5, trans. Bernadotte Perrin, quoted by B. Latour, *We have never...*, 110.
5. Latour, Bruno. *We Have Never...*, 109.
6. Latour, Bruno. *We Have Never...*, 10–11.
7. Laporte, Jean. *Le rationalisme de Descartes*. Paris: PUF, 1950, 220.
8. Noble, David. *The Religion of Technology. The Divinity of Man and the Spirit of Invention.* New York: Alfred A. Knopf, 1997, 143–173.
9. Noble, David. *The Religion...*, 9.
10. Noble, David. *The Religion...*, 206–7.
11. Bernal, John D. *The World, the Flesh and the Devil. An Enquiry into the Future of the Three Enemies of the Rational Soul*. London: Kegan Paul, Trench, Trubner & Co.; New York: E. P. Dutton, 1929.
12. Bernal, John. *The World...*, 73–4.
13. Bernal, John. *The World...*, 95.
14. Latour, Bruno. *La clef de Berlin et autres leçons d'un amateur de sciences*. Paris: La Découverte, 1993.
15. Latour, Bruno. *La clef...*, 66.
16. Fleckenstein, Joachim Otto. "Die Einheit von Technik, Forschung und Philosophie im Wissenschaftsideal des Barock", *Technikgeschichte*, Bd. 32, 1/4, 1965, 22.
17. Fleckenstein, Joachim Otto. "Die Einheit...", *Technikgeschichte*, Bd. 32, 1/4, 1965, 19.
18. Shapin, Steven. "The Invisible Technician", American *Scientist*, Vol. 77, N° 6, 1989, 554–563.
19. Gerland, Ernst (ed.), *Leibnizens und Huygens' Briefwechsel mit Papin*, Berlin: Ferd. Dümmler, 1881.
20. *La vie et les ouvrages de Denis Papin*. Par M. de La Saussaye... et A. Péan terminé par L. de Belenet. Blois: Imprimerie G. Migault, 1893 (1894), Tome Septième et Huitième.
21. *La vie et les ouvrages de Denis Papin*. Par M. de La Saussaye... et A. Péan terminé par L. de Belenet. Blois: Imprimerie G. Migault, 1893 (1894), tome quatrième, 175–178.
22. Ravier, Emile. *Bibiographie des oeuvres de Leibniz*. Paris: F. Alcan, 1937.
23. Amontons, Guillaume, "Moyen de substituer commodément l'action du feu à la force des homes et des chevaux pour mouvoir les machines. 20 juin 1699", in: *Histoire de l'Académie Royale des Sciences*, 1699... Troisième édition... Paris... 1732, 112–126.

24. Ernst Gerland, *Leibnizens…*, 282; *La vie et les ouvrages de Denis Papin*, tome huitième, 144.
25. Ernst Gerland, *Leibnizens…*, 331; *La vie et les ouvrages de Denis Papin*, tome huitième, 210.
26. Ernst Gerland, *Leibnizens…*, 330; *La vie et les ouvrages de Denis Papin*, tome huitième, 209.
27. Ernst Gerland, *Leibnizens…*, 333; *La vie et les ouvrages de Denis Papin*, tome huitième, 212–3.
28. Ernst Gerland, *Leibnizens…*, 340; *La vie et les ouvrages de Denis Papin*, tome huitième, 221.
29. Papin to Leibniz, 11th August 1704. Ernst Gerland, *Leibnizens…*, 329; *La vie et les ouvrages de Denis Papin*, tome huitième, 207.
30. "mais de dire que la matiere en mouvement resiste continuelement à l'Entelechie qui la meut, c'est, ce me semble, donner de l'inclination aux choses qui en sont incapables, et multiplier les êtres sans necessité". [but to say that the matter in motion constantly resists the Entelechy that moves it is, to my mind, to bestow inclination upon things that are not able to have it, and to multiply entities beyond necessity"] Papin to Leibniz, Kassel, 2nd April (23rd March), 1699. LBr (i. e. Leibniz Briefwechsel) 714, f. 314v. This transcription of mine is included in my edition of the complete correspondence and essays between Leibniz and Papin, forthcoming.

Hans Reichenbach and His Techno-Optimism

Arto Siitonen
University of Helsinki

1. Introduction

Techno-optimism is a positive attitude towards technology, technical devices, gadgets, machines, and instruments. It contains the claim that human life is improved by devising continuously more sophisticated techno-systems. One would hardly be surprised, if by a social-psychological questionnaire it were found out that a majority of engineers and the personnel of a technical research institute or school are techno-optimists. In fact, techno-optimism lies at the core of their professional ethics.

Techno-pessimism is the reverse attitude. It need not be motivated by mistrust towards technology, let alone by a Luddist fear of machinery and riot against progress. It need not exemplify technophobia. (On technophobia, cf. Goffi 1988, pp. 5 ff.: "La technophobie intemporelle", and pp. 10 ff.: "La technophobie contemporaire"). Rather, techno-pessimism may also be based on calculations of risks and be fed by environmental concerns. It may also be selective in its concerns and be directed at some special dimensions or branches of technology. (On techno-optimism and techno-pessimism, cf. Tiles & Oberdiek 1995, Ch 1: "Conflicting visions of technology").

The philosopher Hans Reichenbach (1891–1953) can be considered as a representative of techno-optimism. Beginning his studies in 1910 at the Technical Institute (Technische Hochschule) of Stuttgart in the discipline of construction engineering (Bauingenieurwesen), he worked later for the radio industry 1917–20 in Berlin and 1920–26 as assistant of the physicist Ernst Regener in the Stuttgart Institute. There he gave lectures on radio technology, as well as on the theory of relativity. He had interrupted his engineering studies at an early stage – in 1911 – because of his more theoretical interests: he was disappointed because technical facts were not explained in depth by teachers. His new fields of the study were mathematics, physics, and philosophy. These studies he continued in Berlin, München and Göttingen. During the World War I, he served as a telegraphist on the Russian front (cf. Gerner, 1997, p. 10 and 208).

There are not many professional philosophers who have worked in engineering and physics; Reichenbach is one of these. He retained his interest in practical and technical matters throughout his life and formulated his ideas and theories so that these were close to praxis (cf. Gerner, p. 10). Alongside with his studies and academic teaching, he gave popular lectures transmitted by radio; on the basis of these, he published two books: *Vom Kopernikus bis Einstein* (1927) and *Atom und Kosmos* (1930). Not only did he broadcast through radio but also wrote a handbook on this technical apparatus, *Was ist Radio?* (1929).

Reichenbach did not create an explicit philosophy of technology, but his views on the role of techniques in culture can be extracted from his works. As his activities in radio programs and his other pedagogical commitments indicate, he subscribed to the ideals of the enlightenment. Thus, he saw technology as an area that improves our thinking ability and helps us to understand the world – including the imperfections of human perception that science lays bare and that can be rectified by using various instruments. He says in the introduction to his book concerning the radio (1929, p. 9) that the human spirit ("der Menschengeist") manages to improve its deficiencies by creating the art of technique ("die Kunst der Technik"). He also claims that the ideas on the basis of which physicists and engineers have created the new techniques, can be made understandable to lay persons.

Below, Reichenbach's views on technology will be traced from his publications. Critical discussion will complete the presentation.

2. The Radio Book

Reichenbach delivered lectures on radio technology in the Technical Institute of Stuttgart from 1920 onwards. His book on this topic, Was ist Radio?, was published in 1924. In this year he also published an article on the theme of the book, "Radiotechnik und Kultur". The second enlarged edition of the radio book appeared five years later. Here a co-author was added: Dr. Fritz Noack, a connoisseur of radio amateurism. The publisher of both of these editions was Richard Carl Schmidt & Co, in its series "Die Radio-Reihe". This series was edited at the time of the publication of the second edition by "Professor Dr. H. Reichenbach, Berlin".

In the Preface to the second edition Reichenbach tells that since the appearance of the first edition radio technology has gone through such an extensive development that the revision of the book became necessary. At the beginning of the decade, radio apparatuses and radio programs were unknown factors, and one had to advertise the cultural contribution of broadcasting. Today, he says, every child knows what radio is. Nevertheless, the what-question is not answered by mere knowledge of the external outlook of radios. What is required, is a thorough clarification of the physical and technical foundations of radio technology (1929, p. 5). By way of comparison, one may give an actual example. Cellphones are today familiar instruments to everybody; but how many (or few) of us can give an accurate account of the technology in them and in the network that enables their functioning?

The physical basis of broadcasting is given by electric waves. These cannot be perceived by human senses alone, in contrast to light waves and sound waves. However, by the help of radio technology ("die wunderbare Leistung der Technik", Introduction, p. 9), we are able to observe radio waves in the following sense. The technology transforms these waves into electric currents and these in turn into sound waves, and when we hear the sound we are at the same time able to "perceive" – metaphorically speaking – the radio waves as well. One may wonder how it is possible that such waves oscillate in space. Reichenbach claims that his book gives an answer to this question in such a way that it can be understood by everybody. When lay persons are given a presentation of wireless telegraphy and telephony; they can understand the massive technical effort that lies behind these. The book also gives practical hints to those who want to build a radio receiver or improve one. Moreover, the book gives an account of transmitters, short wave traffic between radio amateurs, and technical improvements that are closely related to the radio (Introduction).

The chapters of the book deal with electro-magnetic waves, wave length and range, valves, transmission of language, aerials, receivers, loudspeakers, amplifiers, batteries, and short wave technology. There are also chapters on the organization of a radio station, on broadcasting tradition in various countries, and on the modern applications of radio technology. Among these applications belong the navigation systems of ships and aeroplanes.

When Reichenbach speaks of valves, he presents an historical overview from Heinrich Hertz and his discovery of electro-magnetic waves to Marconi, who transformed this discovery into a technique, and through Poulsen, Fessenden, Arco, and Goldschmidt to Fleming, who introduced valves into the wireless technology (p. 31 f). When he discusses language transmission, Reichenbach presents a philosophical analysis of language – thinking as well as the spoken and heard language. Our ears can read the modulated waves like a book, and radio waves are such a book (p. 44 f).

3. From Copernicus to Einstein

Reichenbach's first book based on his radio lectures – i.e., the lectures that were broadcast through radio – was published in 1927. Its first American edition appeared in 1942. The work deals with the problems of space, time, and motion, as well as with the development of modern astronomy. Its name indicates the two major transformations of science.

According to the author, Einstein's theory of relativity blended old and new knowledge and was derived from many sources. Among these were various measuring instruments, calculations, empirical experiments, and thought experiments. There are eleven figures in the book that document the way of scientific discoveries that led to the theory and helped to test it and make it understandable. These are the following:
A Tycho Brahe's Sextant (p. 17), Roemer's Observation of Jupiter's Moon (p. 26), The Phenomenon of Interference (p. 30), The Total Spectrum (p. 36), The Diagram of Michelson's Experiment (p. 43), A Diagram of the Measurement of the Speed of Light (p. 46), A Diagram of the Course of a Light-Signal (p. 48), Einstein's "Box Experiment" (p. 67), The Curvature of Light-Rays in Einstein's Box (p. 74), The Einstein Tower in Potsdam (p. 77), and Major Segrave's 1,000 Horsepower Auto at Full Speed (1957a, p. 84).

The Einstein Tower was built by the astronomer Erwin Freundlich in order to test retardation of light in the gravitational field of the sun, and by that, the general theory of relativity. Reichenbach says that the tower is "a structure combining to perfection every astronomical and physical contrivance" and gives a detailed characterization of its structure (p. 78). The American translator has added the following note: "The experiments in the Einstein tower could not be continued since Professor Freundlich was forced to leave Germany when the Hitler government came into power." He adds to this sarcastically: "The Einstein tower was given a new name and is now used for purposes which the Nazi government deems less dangerous for the German race" (p. 78 n).

The picture of major Segrave's speed record car is meant to illustrate the relativity of simultaneity. Reichenbach clarifies this by an example of photographing fast-moving objects. Due to the properties of the focal-plane shutter, certain distortions of the picture necessarily occur: the wheels of the automobile "assume the shape of a somewhat crooked ellipse with a forward tilt" (p. 84).

How pure theory can lead to technical innovations, is exemplified by the confirmation of Maxwell's theory by Hertz's experiments:

> These electrical vibrations produced by Heinrich Hertz in his laboratory were nothing other than wireless waves, known today as radio waves. Their widespread technical use in telegraphy and radio constitutes a proof of how a discovery made purely for theoretical reasons, that is, in search of understanding natural phenomena, can yield unsuspected industrial benefit, never thought of even by the discoverer himself. (p. 35 f)

4. Atom and Cosmos

The second book based on Reichenbach's broadcasting activities was published in 1930. The American edition, revised and brought up to date in collaboration with the author, was first published in 1932. The work gives an account of modern physics and concerns the following topics: space and time, light and radiation, matter. The final chapter deals with philosophical consequences. As the name of the work indicates, its scope reaches from the interior of atoms to the interstellar universe.

According to Reichenbach, specialized science has received cultural value for the non-specialists due to its central importance for the fundamental attitude of our age toward life. This interest is of a largely theoretical nature – it arises from the efforts of explaining and comprehending the steadily accumulating experimental findings. Observation is enhanced by various instruments, such as mirrors, telescopes, microscopes, and spectroscopes. Various new results have aroused certain technical hopes; a telling example is the discovery of radiation by Marie Curie in 1898. Thus,

> ... the prophets of the future were full of the most daring dreams of technical progress, which were thus imagined to be realized. People saw before them a future generation in which a little bit of radium salt replaced lamps, in which radium rays penetrated all secret places, and in which all machines and power stations were replaced by a little piece of metallic radium that would furnish all the energy required; the very name, radium, had won a magical sound, and meant for the large public something like the Philosopher's Stone which the middle ages had sought. (1957b, p. 204)

One may hear a strong irony in this remark. The irony is reinforced in the continuing text, in which it is said that of all the expectations "not much more remains to us than the glowing hand of a watch, whose feeble light can at least be seen in deepest darkness" (Ibid.) In conclusion, "aside from certain medicinal curative properties... radium has not yet brought technical progress." (Ibid.)

Although Reichenbach was later to publish the work Experience and Prediction (1938), he certainly did not want to assume the role of a "prophet of the future". He told to the general public of certain scientific innovations and their background in culture, considering technology as a by-product of scientific research. He was as optimistic of technological progress as of the progress of science. Thus, when he speaks of the implications of Einstein's basic equation, he presents the following sober remark:

> If, then, the mass of the atoms could be destroyed, technical sources of energy would thus be created, surpassing everything previously known. It is possible that a technique for using these quantities of energy may later be discovered; for the present, there are advantages in our inability to use them. For it is as yet impossible to foresee the effects of such an exploding mass on the masses surrounding it; the explosion might conceivably be so violent as to involve all neighbouring atoms and their whole vicinity – perhaps our entire planet – in the explosion. We thus live, as Nernst once formulated it, on an island of gun-cotton, for which, thank God, we have not yet found the match that would set it off. Our ordinary matches have only a temperature of about 1,500 C.; we should need about ten thousand million degrees to "set off" the atoms. This difference in scale is, fortunately, still so great that we need not fear the inventions of our physicists in the near future. (p. 208)

The principle that each substance has its peculiar combination of wave lengths, is important for chemistry and its applications: "The knowledge of this characteristic radiation has proved useful for chemical technology, also, since the spectrum given off by an unknown substance can be used for its identification; the spectroscope is, therefore, an important tool of chemical analysis." (p. 226) He adds that spectral research has turned out even more important for theoretical purposes, enabling the construction of "spectral apparatus of ever-increasing size and precision." (Ibid.)

Although chemistry has its conceptual foundation in physics, it maintains "in its scientific technique" its autonomy, "just as electrical technology remains an autonomous science, even though subordinate to Maxwell's physical electrodynamics." (p. 243)

Practical applications of science improve our theoretical understanding of it. Thus, irradiation of various substances with X-rays has produced knowledge "from technical points of view, as to their structure." (p. 246) Causal thinking has received confirmation from practical applications:

> The construction of machines of unexampled technical perfection, which was the practical result of such a science, was at the same time an ever-repeated confirmation of the underlying causal hypothesis, and no engineer would ever attempt to build or repair a machine, using any but a causal point of view. (p. 269)

In these cases, a certain feedback effect is involved; it goes from applications to improved theoretical understanding.

One may note that in the present work, Reichenbach employs not only the word 'technique' but also the word 'technology'. However, by the latter word he does not refer to the whole techno-system or to a subsystem of it, as the word is customarily employed today. Rather, 'technology' is understood by him to refer to a science: "electrical technology", "chemical technology". When he speaks of scientific techniques, he means the procedures and instruments used in research. The German word "Technik", as employed by him in the book on radio, is on the one hand concerned with specific tools and methods, on the other hand embedded into a wider cultural context.

5. The Rise of Scientific Philosophy

This is Reichenbach's best known work, a eulogy to science and to science-inspired philosophy – and to technology. Reichenbach claims that nineteenth-century science contains an abundance of technical discoveries and a wealth of logical analyses, and these developments led to the rise of a new philosophy. He compares this philosophy to the traditional one and finds in the latter many mistakes that have been corrected by logic-driven clear thinking.

The following remark introduces the comparison between the old and the new era:

> Compared with the Renaissance, or the periods of the literary classics in England, France, and Germany, the century of science and industry offers the colorless picture of a civilization striving toward uniformity and mechanization. Mass production instead of creation by the artist or craftsman; mass satisfaction instead of the standards of taste of an intellectual nobility; mental group-work instead of the creative work of the individual thinker – those are the labels which the romantic interpretation of history has in stock for the nineteenth century. (1951, p. 119 f)

Then, a full-blown techno-optimism takes the stage:

> But the history of the age of science and industry will never be understood by the romanticist. The intellectual achievements of the nineteenth century cannot be measured in terms of great personalities – although there are such personalities – because the contribution of an individual, outstanding as it may be, is small compared with the group product. The number of scientific discoveries through group-work during this period is overwhelming. The period that began with the steam engine and the discovery of the electric current, that went on by way of railroad, the electric generator, the radio, the airplane, and that in our day has culminated in supersonic speed of transportation and the tapping of atomic energy – this period in not only a triumphal march of industrial discoveries. It represents, at the same time, a line of rapid progress of the powers of abstract thought. (p. 120)

The industrial civilization is distinguished by the high level of abstraction that is required from education and from the actual working practices: "Never in the history of mankind has a civilization demanded so intensive an intellectual training from those who work for it" (p. 121). Reichenbach takes three professions as examples:

> The engineer who designs machines or airplanes is not identical with the man in the workshop who makes the machine, or the airplane; for him, his product exists completely in his imagination, materialized only in the form of a blueprint, before it can become concrete reality. (p. 120)

This distinguishes engineers from technicians and workmen. Physicists, in turn, stand on a still higher level:

> The physicist who experiments in his laboratory has before him a maze of wires, glass tubes, and metal bars; but into this maze he sees an order of electric circuits,

> which makes him control his manual operations in such a way that observations result which reveal general laws of nature. (p. 120 f)

Physics presupposes mathematical formulas. Consequently:

> The mathematician, equipped with writing paper and fountain pen, arrives at figures which determine the construction of bridges or airplanes or skyscrapers. (p. 121)

On an even higher level of abstraction stays symbolic logic. Because this discipline gives the basic tools to philosophy, one may think that philosophers have the leading role in the province of abstract thought. Indeed, modern philosophy

> ...presents answers understandable only to a mind trained in abstract thought, it requires that its disciples study every item with the precision of the engineer and the scrutiny of the mathematician. (p. 121)

Moreover: "The construction of symbolic logic has become one of the outstanding characteristics of scientific philosophy." (p. 215); "Logic is the technical part of philosophy; for that very reason, it is indispensable for the philosopher." (p. 228) In a way that recalls to the mind of a present-day reader certain recent developments in the areas of computer science and cognitive psychology, Reichenbach envisions the possibilities of progress:

> It may even be used to transcribe logical problems in such a way that they can be "fed" into an electronic computer; and a modern robot of this kind might some day very well be able to solve logical problems which would defy the capacities of the human brain, in the same way that it already solves mathematical problems. (p. 227)

Concerning robots, Reichenbach treats the possibility that is familiar from science-fiction: "Imagine that scientists had succeed in constructing a perfect robot." (p. 271) He then raises the question: "How do you know it has no mind?" (Ibid.)

6. Evaluation

Reichenbach had rather a philosopher's and scientist's attitude towards technology than that of an engineer, despite the fact that he had worked as a radio engineer. Although technically skilful himself, he placed theory on a higher position than practice. In his works, stress is laid on knowledge and understanding, and various techniques serve the purposes of pure research. Nevertheless, he was enthusiastic of technical innovations and praised these even by employing poetic expressions. He considered technology as an art: in his book on the radio, he employs the expression "Kunst der Technik". One may add to this that various technological structures and milieus have aesthetically valuable qualities. As to the Einstein Tower, in today's Potsdam it is considered as an architectonic treasure.

Reichenbach praises technological progress and sees technology as an essential factor in our efforts of improving the human life. Due to these attitudes, he is an outspoken techno-optimist. His position avoids the extremities of techno-romanticism and technocracy. One may find in his texts formulations that suggest both of these utmost views, but his style is marked by an ironical tone that creates distance to them. Irony even moderates the optimistic position itself.

Reichenbach uses the expression 'technique' in two senses. Firstly, there are techniques within scientific research, purely mental ones such as the formalisms in logic and mathematics, or physical ones as the instruments used in research. Secondly, he sees technique as embedded into a wider cultural context, as his book on radio indicates, or even as a culture-forming factor, as his view on industrial civilization in the 1951 book shows.

A critical comment may be presented to Reichenbach's distinction between the two cultures – i.e., the artistic-literary and the scientific-industrial one. Doesn't it give a too polarized picture of social development? Literary intellectuals need not subscribe to "the romantic interpretation of history"; some of them may even excel in the areas of technology and industry. On the other hand, engineers, physicists, mathematicians and logicians may develop fine artistic and literary capacities. One proof of this is delivered by the production of Reichenbach himself – it is marked by a good literary style.

Literature

Gerner, K. 1997. *Hans Reichenbach – sein Leben und Wirken, eine wissenschaftliche Biographie*. Phoebe-Autorenpress, Verlag unabhängiger Autoren, Osnabrueck.

Goffi, J.-Y. 1988. *La philosophie de la technique*, Presses Universitaires de France, Paris.

Reichenbach, H., Noack, F. 1929. *Was ist Radio?* Richard Carl Schmidt & Co., Berlin.

Reichenbach, H. 1957a. *From Copernicus to Einstein*. Translated by Ralph B. Winn. Philosophical Library, Inc., New York. (Originally in German in 1927, first American edition in 1942).

Reichenbach, H. 1957b. *Atom and Cosmos. The World of Modern Physics*. Translated by Edward S. Allen. George Braziller, Inc., New York. (Originally in German in 1930, first American edition in 1933).
Reichenbach, H. 1951. *The Rise of Scientific Philosophy*. University of California Press, Berkeley and Los Angeles.
Tiles, M., Oberdiek, H. 1995. *Living in a technological culture*. Routledge, London and New York.

PART THREE
The World of Technology and Its Ethics

Human and Machine Centred Technologies: A Philosophical and Ethical Analysis[1]

Timo Airaksinen
and
Ville Paukkonen
University of Helsinki
Finland

1. Introduction

Traditional humanistic ethics of technology may well be based on a false assumption. It seems to evaluate all technological devices and systems from a human-centred point of view. For example, if technology is experience it must be human centred (McCarthy and Wright, 2004, pp. 12 ff.). Experience is an anthropocentric notion. But not all technology is human centred nor should it be. There is and will be resistance to technology exactly for this reason (see Bauer, 1997, pp. 19 ff.). People know instinctively and pre-theoretically that not all technology is for them. Traditional ethics of technology may not, for this reason, give us the necessary ethical tools we need to evaluate the different technologies. It is simply too general and vague and, thus, trivial in its best case and misleading in its worst. We intend to show four relevant aspects of evaluating technology ethically which play an important part in practically all of the devices and apparatuses we encounter. But, what is even worse than biased or trivial ethics of technology is no ethics of technology at all; this situation is properly called "the silence of technology." We shall argue that and explain why, because of the inherent aspects of technology, policy makers often fall victim to this situation. It is against these two, humanistic ethics of technology and the lack of ethics of technology, which this article argues. In addition, we will attempt to pose questions which any viable future theory of ethics of technology should recognize and at least partly answer.

2. Technology as a Network and the Forms of Technical Action

First of all, it should be noted that a given technology is not a collection of material devices and apparatuses. That collection is only a part of the technology, which is *a resource* for our actions, as it is in Praxiology. Unlike a premodern tool, like a hammer, any modern technological device is mere junk without its supporting network. A good example of this kind of device is an ordinary toaster. What do you do with a toaster without electricity, spare parts, and cut bread? Not to mention all the raw materials and transportation networks which are already presupposed in the example? The answer is: simply nothing. The reason we fail to notice this and think of technology as the constellation of independent objects is that our paradigmatic examples of technology are from the premodern period; such as hammers and other simple tools which require nothing but individual skill, muscle power, and easily attainable materials. But, the paradigmatic "tool" of our age is, by no means, a hammer but rather a toaster or some other of the many common white goods. Our primordial material needs are satisfied by some technological systems almost too large to be comprehended by the individual human mind; such as modern agriculture or traffic controls. What are left forth are the numerous luxuries that make our life meaningful and happy.

Technology is a praxiological resource that enables us to do things never imagined before and enhances our natural abilities, but it also turns us into a resource among our technological devices. This is not to be taken in any Heideggerian sense, implying that we would turn completely alike our technological objects of rational calculations, but merely as a thesis that our actions are not only made possible by modern technology but that they also make modern technology possible. Whenever I decide to buy eggs at the supermarket rather than to grow my own chickens and use their eggs, I validate the huge technological network called agriculture. And since I do it often, I am confined to the modern technology, no matter how critical my cognitive attitudes against it may be. But, our independence from technology dissolves our capacity to act. Instead of spending my days feeding my chickens, I can read a book or go to the movies or do whatever it may be that I desire. We are inclined to say that technology decreases our independence but increases our autonomy. But this applies only to those who have an access to and are able to dispose of modern technology, namely citizens of rich western countries. This does not mean that arguments put forward in this paper would be fallacious in relation to the problems concerning technology in a so called "third world." More work is required to analyse the situations where technology has not yet completely penetrated society. We shall deal in this paper with the situation where practically all developed technology is in use; namely situations in any western country. Eventually, we argue hopefully, this will most likely be the situation in developing countries.

Having said something about the concept of technology, we now turn to the question of the differences within the realm of technological objects and the ethical consequences they imply. The following figure is essential for our arguments. We shall deal with all of its parts one by one in order to convince the reader of its importance in the decision making related to the ethical questions concerning technology. In each cell we will give an example, and inside the brackets we will give a theoretical term which provides a context for the example. For instance, safety is something we desire whereas welfare belongs to the category of needs.

Figure 1

The purpose of technology:	Human environment: CENTRIFUGAL	Human environment: CENTRIPETAL
OBJECT CENTRED	Safety (Desire)	Welfare (Need)
HUMAN CENTRED	Automation (Function)	Taylorism (Power)

3. Human Centred and Centrifugal Technology

The purpose of *human centred* technology is easy to understand. By being humans, we understand the concept of a human agent at its value-bound, ethical meaning; humans as an end as such, not as a means to something. This is truly humanistic, praxiologically viable technology; the everlasting dream of many cultural critics like Herbert Marcuse or Lewis Mumford, just to mention two. Human centred technology is made for humans. It functions according to human values and respects the human person and its uniqueness.

A centrifugal environment means a situation where the critical gaze is moved from the centre; a situation where we submit ourselves to a process that leads us gradually away from the centre. Our example is *safety* as a value and as an objective. First of all, safety is always human (or animal) safety. We do not care about the hard luck of machines in any moral sense. We may grieve for our economic losses, but we do not or we should not feel any empathy for the broken machines, which is unlike what we should feel for humans. But, safety is also centrifugal. It is a value that should come true, but at the same time we do not want to get into a situation that involves safety risks. Safety is understood as the *absence* of risk. Safety implies a world without something, risk. In that sense it is a centrifugal notion.

Risk is a concept with many meanings (see Gigerenzer, 2002). Most of the time we identify risks with an epistemological uncertainty about certain events. Some human value has to come true in a situation which is now threatened by a sudden change. The change in process does not need to be a major one, just enough to cause a negative change to the instantiation of the appreciated human values. This change does not need to cause a catastrophe from the point of view of those human values; say switching off the power to a computer that has some important, unsaved documents on it. We must anticipate these new situations so that we can prepare ourselves, or we try to eliminate the risks; say put auto-save on or make lots of backups. But, very often, the risks are not foreseeable. On the other hand, we get accustomed to some of the risks to the extent that they do not feel like risks anymore, but this is an irrational attitude towards them, for we should be afraid of something that is dangerous.

We should think about centrifugal environment as a situation where we attempt to describe, find, evaluate, and expel some of the causes for the risks from the centre. This way we can feel safe again. This is what we desire. The problem is that although it is often easy to describe the risks, it is often just as hard to actually locate the risks before they come true, and then, of course, it is too late. Science fiction is full of horror stories of what could go wrong. The reduction and elimination of the risks takes skill, will, money, and resources. Moreover, sometimes we even, willingly, seek the risks as in extreme sports. Risks are not always something to avoid. Humans express their freedom by turning a centrifugal environment into a centripetal environment. Perhaps playing with the risks gives us the feeling of being alive, of power, and maybe even of immortality. Still, in rational thinking, risks are something to avoid. Technology solves this contradiction by enabling us to simulate the risks and at the same time maintain a factually risk-free situation. This way we may feel an important experience of the risk without any actual risk being involved. The best examples of this kind of simulation are probably television programs that create the risks in the studio in order to give some dramatic tension to some otherwise boring programs. The TV-viewer identifies his/herself with the actors from his home couch and gets to feel the tension and excitement without any immediate physical danger.

The safety of technology is one of the three basic values of an engineer's professional ethics in addition to his health and well-being. However, *safe technology* is not the same thing as *safety technology*. The former, safe technology, is any technology that contains as small a risk of an accident as possible. The latter, safety technology, is any technology that is safe to use and, in addition, is meant to eliminate the risks of an accident. As an example, a roll bar on a tractor is a safety technology that is meant to protect the driver in case the tractor rolls over. Such a roll bar is a cost factor with no other purpose than to protect the driver in an accident, i.e. a situation where a normal situation turns into a catastrophe; a small error in driving a tractor leads into a situation where the human is crushed between earth and metal.

Safety technology is often passive. Active safety technology is hard to conceptualize. An assault rifle is safety technology when used by someone with skill. The shooter secures his safety by using a rifle. At the same time other people nearby are exposed to the risk of getting injured. In this case, security is attained by multiplying the risks. The difference in passive safety technology is that we do not need to speak or think about passive safety technology, whilst active safety technology forces us to discuss it constantly. Only, in that way, may we recognize the risks it abolishes and the new risks it possibly produces. This is a difficult situation in a centrifugal environment where the production of risks is prohibited. The whole idea is to get rid of the risks. This is especially clear when discussing surveillance technologies. They may serve our security and, thus, be centrifugal (say surveillance cameras in banks), but it may also turn into a servant of business and profit in which case its centrifugal nature disappears. Besides this, surveillance technology is rarely human centred but is often straightforwardly alienating. All safety technology should be unambiguously human centred and centrifugal. This is the only way of keeping it under discussion, and the only way of supervising the ones who monitor us.

Other possible examples of human centred, centrifugal technologies are the many medical technologies. Technology that is designed to eliminate maladies is clearly centrifugal. On the other hand, this is too simple as a picture of illnesses. The modern way of thinking about medical technologies is often health-centred; health is a value that should be brought as close to the person as possible. Instead of getting rid of an illness we focus on our renewed vitality, activity, and happiness – this is health. And this kind of health exists in a centripetal environment. Instead of speaking about illnesses (which is a loss) we tend to speak about healthiness (which is a gain). Thus, medical technologies are sometimes centrifugal (dispelling sicknesses) and sometimes centripetal (inducing health). Yet they are always human cantered (see Pacey, 1999).

Our last and most interesting example of human centred and centrifugal technology is the technological improvement of humans. This is the post-humanistic dream: the idea of a cyborg is to replace weak human organs with superior technological improvements. A cyborg is clearly the product of a centrifugal environment in which case the aim is to eliminate human deficiencies. It is also human centred, supposing that the creature created is still human in some common sense of the word. It may be that it makes no sense to speak of a cyborg as human, in which case we would not be dealing with human centred technology (see Brooks, 2002 and Clark, 2003).

Post-humanism is ethically ambiguous (see Hayles, 1999). If we start to think of a human as a British Museum of faults and deficiencies, as Mark Twain once put it, we are obliged to take the negation of the human element seriously. We admit that the human is simply too incomplete to be of any value as such, and that there is nothing in the human that could not be replaced or improved on by a machine better than the original bio organ; if not now then at least in the future. On the one hand, we lose the human if we replace all his bio parts with technology, whilst on the other hand, all is done for the human's sake. The same problem is apparent in our speech about new information technology, ICT. If we affirm this technology we do not seem to care about humans, but if we reject these new technologies we forbid humans the right to develop themselves and improve their current situation. Even if we admit that the new cyber technology is not human centred, we are not necessarily facing a scandal. Many good technologies are object centred.

4. Human Centred and Centripetal Technology

Well-being is a value that technology, for its part, produces and should produce. It is closely related to human needs. Without it human life is not complete. Technology produces two kinds of well-being: *hedonistic* and *functional*. Hedonism is pleasure and enjoyment. Functional well being means the good quality of an instrument. The well-being that technology produces is direct if technology produces pleasure or functions as an instrument. It is indirect if it produces something that can be used for pleasure or which leads into the accessibility of good instruments. In a centripetal environment, technology produces something that moves into the centre of human life. Understood in its broadest meaning, well-being is the only concept we need in this instance; only technologically produced well-being is both human-centred and centripetal.

The well-being produced by technology is hedonistic when machines and their systems produce pleasure for their users (see Oudshoorn and Pinch, 2003). A hot bath is a good example. The difference between a hot-water tank and the hedonistic pleasure that the human gets from it is minimal. Hot bath is a form of pleasure. A refrigerator, on the other hand, gives us a functional well-being. Its primary function is to keep the food tasty and to prevent it from going bad. It is a good tool in the service of well-being. We rarely appreciate it like we do with direct well-being except, of course, when the refrigerator breaks down.

One species of well-being is not produced by technology. This is emotive well-being. There is no direct relationship between machine and emotions. Machines are just machine and not something else. Although people may and do, in fact, fall in love with their machines, this is not because of the machines themselves but because of the humans who project their wishes and desires on to the mechanical objects. Thus, this sort of attachment is fetishist by nature. One example of this may be a robot dog named Aibo that Sony manufactures. This is not to say that keeping any pet would not be a sign of an emotive void, just that keeping a robot dog is so in particular. But, if technology does not produce emotional well-being by itself, people are in danger of facing serious emotional problems in a completely technological world, or more precisely, facing a shortage of emotions. This would negate the very idea of human centred technology. We need our emotions.

The main reason why people project feelings and emotions on to cold artefacts is, of course, the advertisement industry, which persistently assures us that by purchasing, for example, a Mercedes car, I not only get an automobile with a powerful engine, I also get sexual satisfaction, interest from the opposite sex, and respect from other motorists. And of course, similar "arguments" are used when marketing white goods and cosmetic products to women. The whole idea of such an advertisement is to dissolve the difference between a machine and an emotion to a minimum. When only a few people have automobiles they signify success and arouse envy in other people, but when everyone has one, then to believe an advertiser's promises is to live in a dream. The belief in it is based on the memory of a time when an automobile truly meant being something more than just a common instrument. But promises based on memory never last long and the joys of purchasing a new car are soon followed by uneasiness and disillusionment. A new car isn't really anything more than just an instrument and, as such, withdraws from the sphere of meanings into the silence of technology.

The problem that humans face is that the system the machines form is not merely functionalistic, it is also machine-centred. Machines colonialize their environment and constantly threaten to move beyond the centripetal borders of the human needs and well-being. In the end, the human only encounters machines, communicates with them, is resourced by them, and then learns only those skills that the utilization of the machines requires. In this case, our environment, "lifeworld" if you like, loses the meanings it normally carries with it. Deprived from facing his and others' emotions, the human starts to fear genuine human dealings and interactions, and because technology produces well-being, which is good-in-itself, there is no way of going back to that old state before technology. Indeed, the well-being that machines produce is centripetal, but it also contains the possibility of, or rather the danger of, a centrifugal situation. We are going to loose something valuable.

For the well-being of a centripetal environment that technology produces to be possible and complete, we need emotions. They are acquired, among others, from entertainment, which is industrially produced and broadcasted by machines. In this special case, the machine does, in fact, produce a functionally emotional well-being, but only in an indirect fashion. Television, for example, brings moving pictures and loud sounds to our homes. Part of this material is entertainment and part of it produces emotions. Thus, television produces emotions, but indirectly of course, because the actual function of television is to produce moving pictures and loud sounds, whereas part of these moving pictures, either the entertainment or the news, is designed to produce emotions. The mere functioning of technology is not enough, and only a few can love television as an apparatus. What are needed are meanings. Everyone either loves or hates television programmes and, thus, attaches emotions to them. The so called "content providing" is exactly this, which is to provide the content for the different mediums which are mere forms. Television programmes are produced so that people will buy more televisions, and computer programs are produced so that people will upgrade their old computers. This is all part of McLuhan's old finding that "medium is the message." The difference between an instrument's function and its indirect effects is not as great as is usually presupposed.

The term "entertainment industry" is not an industry only metaphorically, but is an industry in the very real sense. Industrially produced entertainment means that technology produces emotions in us so that we can use them to replace the lack of emotions that a technological life itself leads us into, and there is no going back. Well-being is thus defined by technology and no one is willing to give it up. There is no reason for sentimentality. Every one realizes that it is not reasonable to play *Wheel of Fortune* or any other quiz show all the time, at least not so much that one would take an interest in it day after day, but when it comes to watching the same thing from a television the situation is completely reversed. And no one finds this curious. A human is willing to voluntarily regress into a childlike behaviour so that he or she will be able to feel the emotions the quiz shows offer. And he has no other chance; otherwise he is left completely without emotions in a technological world where the traditional way of expressing and receiving emotions from interaction with family and friends is quickly transforming into different machine based models of communication. However, the emotions we recognize from television are rather primitive; mainly excitement from watching violence or adolescent erotic feelings from soap operas.

One possible explanation for the poor level of emotions produced by television is in its nature as a one-way channel, without any possibility of feed-back. This is not the case in genuine social communication, where we, willingly or unwillingly, regulate our and others' emotions which in turn guarantees that they will not stop at the surface level where they are easily recognisable and often banal. Instead, we develop and refine our emotions in social communication situations, especially when the relationship between different participants of communication is free and sustainable. The technological mediums do not allow this, but they do reinforce the emotions that are easy to produce in the television studios, and which are then communicated to as many people as possible; in other words, as simple and fundamental emotions as possible. As we said before, these two basic emotions that entertainment deals with are, as anyone can easily perceive, immature erotic love and straight violence. The previous arouses and the latter shocks. The spectator is constantly aware that she is dealing with a fantasy that does not threaten her own existence, or for that matter does not excite any contemplation of it. Everything in this system is designed to reinforce this illusion. Performers are picked out so that their faces will match their roles. The evil ones are ugly, of course. Cosmetic surgery and makeup put the final touches on an appearance that is both imaginary and plastic. Acting does not correspond to normal human psychology but it does follow in its own ritualistic laws.

The key word here is *repetition*. The machine makes endless repetition possible, which in turn produces habit and indifference. Repetition alienates us from the subject in question and decreases our emotional attachment to the situation. That is the reason why news from different catastrophes around the world moves us very little; images loose their original strength after constant repetition. A formulaic narrative has the same effect. It allows the spectators to foresee what is going to happen and, thus, prepares them for the emotion they are about to experience. This way, the emotion is under control and neutralised. The absolute threshold can be increased, which makes even more cloying love and barbarous violence possible without any backlash. The rejection of openly sexual and pornographic content, especially in the American cinema, is merely an exotic cultural prejudice that shall soon be overcome. Entertainment increases the intensity of the stimulation but produces less and less emotions in the audience. The same logic of repetition not only governs entertainment, it governs the news as well. The news broadcasts create well-being, not because they would give us a truthful picture of the world but because it reassures us and creates the feeling of safety. A day in the news starts with an alarming incident, which is then repeated for the rest of the day until our calm returns. The first instance of news awakens a welter of emotions; fear, pity hope, etc. These emotions are based on our ability to identify with the victims of a current catastrophe. We feel insecure, even though we are, in fact, sitting safely on our home coach and forced to follow "the story" to its end. After a few repetitions our emotions start to reduce down from their original authenticity until there is only a faint memory of them left. The same pattern goes on every day until the human learns that there is no danger. Although the morning news may shock us, by the evening our feelings are calm again. The media's way of standardising all the news to resemble each other works in favour of ourselves feeling safe and in control of the situation; new day, same old news.

As long as we need emotions as an integral part of our well-being, we are doomed to misery. The machines just cannot fulfil that need completely or even satisfactory. After all, there is a void in the centre of human centred and centripetal technology; the lack of authentic emotions. Notice, incidentally, that many people think they can live without emotions, that is, they do not desire them. But they need them in the sense that their mental life suffers without emotions. The post-human's wish to replace humans with machines is a logical consequence of the panic that the artificiality of emotions causes. If a human can not be without emotions, at least a machine can. Still, it is clear that modern technology produces enough pleasing emotions that resemble genuine, authentic human emotions, that we can forget about the machines and live our lives without thinking about them; by merely suffering from the mysterious feeling of deprivation. It is possible to say that industrially produced emotions that are transferred through machines are genuinely human emotions, for they are produced and offered to others by humans themselves. Maybe some day they will gain the same complexity and meaning as our original, socially produced emotions once had.

5. Object Centred and Centripetal Technology

Object centred and centripetal technology keeps us at the centre of the attention and action but yet functions on the machine's own terms and has no respect for the special nature and status of the human. Everything is objectified. In this case the human is pulled into the centre of the world for machines and this makes him an object, a component of the technological system.

When we forget about the humans in the situation where the humans, in fact, are at the centre of things, ethical problems are bound to rise up. An example of a morally reprehensible object centred and centripetal technology is the use of slaves. A slave is a person whose identity is altered, on the basis of the law, into an object that can be used like a machine. The slave is an object which can be traded for a new one. The only restrictions to his (mis)treatment are those of economic risk and the decreasing of efficiency; if a slave is mistreated, his work contribution decreases and the owner's profit lessens. In a curious way, slavery represents technology where the slave is both a technical object and a human person. But, this is all metaphorical and meant as an example of the potential ethical problems of object centred and centripetal technology. The slave is not a genuine part of the technological system.

In a modern technological system, an example of object centred and centripetal technology can be found in Fordism and Taylorism. Their human is conceived as an object and as part of a machine. Christian Parenti summarizes Taylorism as a system of work management and rationalisation based on three principles and stages:

1. Information gathering and internalising; where the supervisors of work gather all of the possible information about the work process in hand and discerns it as rules, regulations and formulas.
2. The separation of the mental and physical aspects of work; where the physical aspect is left to the workers but the mental aspect is acquired by management.
3. Work is controlled by information; where the work is planned on beforehand and the worker gets accurate instructions of what to do at any moment.

(Parenti, 2003, 136).

Every work assignment is divided into parts, and the efficiency of performing these parts is carefully inspected. The work consists of the body's movements, and everyone performs his own part in complete isolation from the others. But still or more precisely, just because of that the most efficient outcome is achieved. The assembly line is a typical environment where this method of work management pays off. The down side is that the worker's own skills, knowledge, and control over his work disappear in the process. The only knowledge he has is in his own part of the manufacturing process, which is simple enough that he can be replaced with someone else if he disobeys. For, if the worker knows more about the work at hand than management, he has a powerful asset at his disposal when discussing wages or working conditions. Now that robots have replaced humans on the assembly line, (except in the clothing industry in the Far-East) Taylorism has been forced to transform itself into a digital period. Computers enable more control and power over machines than any manager could have ever dreamed of having in the beginning of the 20th century; and what's more, Taylorian work management can be brought into the offices as well, instead of limiting it to the simplest manual labour.

Taylorism is a form of social power. In Taylorism, management seizes the work process intellectually. Everything is seen as a zero-sum game where the information gathered by management means the lack of the worker's information, and vice versa. This is not common in the world of information where knowledge is supposed to be freely available and utilisable. For example, in the scientific community relevant knowledge is imagined to be at everyone's disposal, at least ideally. At least, this is the dream we have about the information we use to construct beautiful visions about the future of the so called "information society." This is, however, not the case in an object centred world of technology.

In Taylorism, work is not a self-expression of humans, but it is understood as a metaphor of the machine's functioning. The worker functions as a machine without understanding the meaning of his work and for not taking responsibility for the consequences of his actions. In Taylorism, the idea is that the work would not be that effective if the workers had all of the information about their work, because then the possibility would exist of the workers influencing their own work. This situation is object centred and becomes possible only when the machines have thoroughly been accepted and used. In this situation, the work starts to be conceived as something that should be performed with the machines and according to the specific "logic" of the machines. The machine becomes the ideal for the worker and there is no political will or even the possibility of retrieving information from this system of machines. The human is drawn into the centre of this system so that his own capacities, creativity, and values as such, do not affect the functioning of the system by any means. The human is understood as to be replacing some part of the machine in the system. This situation is, at the same time, horrifying and liberating. It is horrifying because the work loses its meaning as the main purpose for life; as an essential feature defining human beings. It is liberating because, even though the worker is obligated to work eight hours a day, he is, for the rest of the time, free to develop his capacities and fulfil his wants as he wishes, mainly because of the material well-being created by the mechanical, inhuman work that he is subjected to. The critical question is whether or not this is possible. Could humans really switch off their "essential being" for half-a-day and then return to it as if nothing happened?

This question was discussed in the context of Marxist thoughts in the early part of the 20th century, but has since been largely forgotten or labelled as a mere negative reaction to Fordism. Marx gives arguments for both parties. On the one hand, working in a factory is alienating regardless of a capitalist economy because the end product is too far from the individual worker on the production line. On the other hand, Marx does not even discuss the possibility of a qualitative change in mechanical industrial work, merely its quantitative decrease in a future communist society. Some later Marxist, namely the so called Frankfurt School and especially Herbert Marcuse on his *One Dimensional Man*, interpreted Marx by saying that alienating work is always condemnable and then widened this as a critique of the whole technical reasoning, "instrumental reason" as they called it. Others, however, were more positive towards the rationalization of work and Fordism in general. Antonio Gramsci, for example, was enthusiastic about the total rationalization of the work process; seeing it as an advocate for the workers caused by increasing their wages and promoting the overall rationalization of society. He went so far as to promote prohibition and workplace surveillance in order to get workers accustomed to the new rational work and society (Gramsci 1971, 304). The question of industrial work, as such, has not disappeared, although the mechanical drudgery has largely moved from the assembly lines of factories to the call centres and supermarkets and such; like the service sector jobs. Most of the people in industrial societies have jobs which pose little meaning to them, whilst discussion about the problem has died. Perhaps this is an indication that Gramsci was right in the first place: working in a non-fulfilling job is a part of the human fate that we should accept as a necessary condition of our being. However, public discussion is not interested in this kind of objectifying work in which the majority of people still occupy. Talk of it is replaced by more fashionable theories of a "creative class" and such.

6. Object Centred and Centrifugal Environments

In modern life humans are not always needed or welcomed. Machine excludes us. They reject us and expel us. Machines are self-sufficient in such a world of objects. If humans can be replaced by robots, we would be discussing centrifugal environments. Since humans do employ positions in the midst of machines and systems, the case is centripetal. But, since everyone seems to agree, in principle, that humans should be replaced by machines in tasks that are too dirty or stressful for humans to handle, the technical environment in question is, in a way, between centripetal and centrifugal. It is *de facto* centripetal since humans do operate in positions that could be operated by machines. It is *de jure* centrifugal since these positions *should* be operated by machines, not by humans. Humans are in a position they should not be in and they want out. Maybe all object centred centripetal technology is only temporal and will be replaced by a centrifugal system in the future. Note that from the workers point of view the question is not that simple. It is unfortunate to be treated as an object in a centripetal system, but it is not exactly a stroke of good fortune to be unemployed either, which would be the case in a centrifugal system where he or she is not needed. But, this is not necessarily an ethical problem, but more of a factual one. From an ethical point of view, object centred centripetal technology is an environment in which the human should not be in, for it does not promote a good life for him or her. From a technological viewpoint the conclusion is the same. If the human is weaker and clumsier than the technology, at least he or she will be, and thus, should be, replaced by a better technology. In a most fortunate case not only did the human free himself from an oppressing technology, but the technology also got rid of the non-functional human.

Object centred technology is not only a blatant misfortune for humans, it also has a huge potential. It is misleading to promote and seek human centred technologies and never object centred ones. There are situations when humans are forced to employ an object centred technology just to make things happen. Crisis situations may offer good examples of these situations. Let's take a hospital emergency room for our example. As long as the patient is unconscious, doctors and nurses treat him or her as an object in this system of complex medical technologies that surround him, thus keeping him alive. But at the minute the patient wakes up, nurses and doctors ought to start treating him as a human being and as an end as such; using the human centred medical technologies they have at their disposal, of course. The situation where the medical staff is "fixing" or "repairing" a patient as an object changes into a situation where they should be nursing him as a person. Thus, the same process (treating a patient) may vary between object centred and human centred, but is still trying to achieve the same goal which is human well-being. Object centred technology is never ideal as long as humans are in the picture, although it may still be necessary in some cases; enabling a higher efficiency and taking care of things that need to be taken care of.

Technology is not just a collection of neutral means to some independently chosen goals. As humans use technology they are, as well, "used" by technology. This is what object-centred technology means, and the only way out of this situation is to expel the machines into their own closed world of automation. But before this happens, the human is forced to live with the machines in a centripetal environment, which is never an ideal situation. It reifies us.

7. The Meaning of Technology

To see the way from primitive tools to the large socio-technical systems is also to see how the number of meanings originally related to technology diminishes through time. It is useful to think this through the hermeneutical concept of "horizon." According to Gadamer: "The horizon is the range of vision that includes everything that can be seen from a particular vantage point." (Gadamer, 1975, 301). The narrower our horizon is, the greater the change is of over and underestimating the things in our field of understanding. Our understanding of an object sharpens and gets more accurate as we expand our horizon and even open up new horizons. A tool's way of being is born into its horizon of meaning, whether it is work, war, or art. In the case of technology, the more primordial the object in question is, the more horizons it opens. This is the case, especially with tools; a hammer may be a tool, a weapon, an artwork or a relic. By placing ourselves into the situation as the user of that object, we may have seen the multiple functions the object may have had. Each of these horizons enables the tool to be perceived as itself, i.e. they constitute its being. Heidegger thought that only one of these differing horizons is technological horizon, but this kind of narrow interpretation does not do justice to the hammers' being in all of its diversity. The hammer presents itself in all of its horizons of meaning, although its being is not in every one of them "being-in-hand," as Heidegger thought.

But as we move forward, our technical objects loose this versatility. An automated hammer in the factory serves only one purpose. Although it may seem that some modern technologies do, in fact, carry multiple functions, such as mobile phones that have digital cameras integrated into them, they do so only in a very limited sense; they are in a way a carrier of multiple tools, in this case cameras and phones. But, they are unable to cross the different horizons as the primitive technologies did. They are limited to purely technical functions. According to Heidegger, this horizon of modern technologies is *enframing*; within it everything shows itself as a means to something outside itself and the human is threatened to be sucked into this technological way of being. Heidegger calls the human's change of position with modern technology as standing-reserve (Heidegger, 1977, 19). This refers to the fact that, according to Heidegger, it is with modern technology that humans, as well as nature, are seen merely as a reserve waiting to be put to use, instead of objects with some intrinsic value in themselves.

As we said before, modern machines are not as flexible between the different horizons as are tools. The relationship between machines and the human life-world is not as intimate as it is between tools and the human life-world. The machines are constantly on the move toward their own object centred machine-world where the human is reduced to a technician that maintains the system. The human is being pulled towards the world of machines which he is even less able to control than the world of tools. There are two sorts of machines that surround us: we shall call them "use-machines" and "industrial-machines". Use machines are something that we spend our time with, like cars and washing machines. These machines surround us at the phenomenological level; we constantly use them and discuss things about them. However, the industrial-machines operate at a level that hides itself from our eyes and from our speech. They are the machines that provide us with the endlessness of use-machines but do so inside the factories that do not interest us. In a way, the industrial-machines are sort of "ur-machines" since they are mainly responsible for all the visible technology surrounding us; machines are already making other machines. Paradoxically, industrial-machines seem to have more horizons of meanings than the use-machines.

For example, a paper-mill may be used for, besides producing paper, teaching and instruction, and finally with any luck, it may end up in some museum of technology. But when it comes to the use-machines, say cars, there seems to be a rather limited horizon of meanings since they are mainly for driving around in or competing with someone in. This is paradoxical because when it comes to symbolic meanings, a car outweighs a paper-mill anytime, and there are all the usual metaphorical meanings attached to cars such as power and sexuality, whilst none of these can be truly attached to a paper-mill. These associated values do not have to do with any horizon of meanings, but they do refer to the metaphorical being of the artefact. Sometimes it may be more important than the actual meaning, which is the use of the artefact, as is the situation in many cases with cars. Whoever is buying a new car is faced with technically identical machines, with the only differing thing between them being the meanings that the car dealers and the market men attempt to attach to them.

To explicate the difference with a metaphor from linguistics, one should think of horizon of meanings as having to do with the *denotation* of the word, or in this case the machine as the thing it refers to; its use or its function. The metaphorical meaning has its equivalent in the *connotation*, i.e. all of the other words that have a connection to the word in question, or in our case all the metaphors and cultural meanings that have a relation to the machine in question and which constitute, at least, its social meaning; such as power and sexual appeal when it comes to the Ferraris and such. It is through these "secondary" meanings (the metaphors that marketing exploits) that constantly brings use-technologies into our eyes and into our discussions. Thus, the truth is not to be found in the use-technologies as they present themselves to us at this moment, but it is to be found in the industrial-technologies that make all the use-technologies possible in the first place. By examining these technologies we come (through speech about technology) to the conclusion that technology constantly escapes speech and critical inquiry; technology's purpose is to be used, or as the situation is today, to function alone and not be discussed. Humans think of technology in an anthropocentric manner, and because people require recognition to exist, we conclude that machines, too, should be discussed either in a positive or a negative tone. But the truth is, machines just want to accomplish their tasks and be left alone in their machine-world. As we do not discuss the tools we use as long as they work, neither do we discuss the machines until something goes wrong, and when that happens all we can do is rely on the machines to fix the situation.

The curious situation of use-machines is centripetal since it drags humans into the centre as the owners, worshippers, and users of technology. At the same time it is object centred since it attaches different cultural meanings to the technology that it does not originally have. This is the situation that the traditional Marxist critique of consumption has had its eye on; the discussions about *fetishism*. Undeniably, this is an important and interesting aspect of technology; humans attaching human values to inhuman objects. But, as we should have seen so far, it is only one level of human-technology interaction and by no means the most important one. Just think of the primacy of industrial-machines over use-machines that are the subject of analysis in any discussion on fetishism. Most of the technology that surrounds us and makes our living possible consists of industrial-machines that escape scrutiny. These sorts of technologies do not address us in any other horizon of meanings other than in their own horizon of production, and it is not a task for the philosopher to discuss this horizon but it is a task for the technician to discuss.

The whole idea of horizon of meaning seems to lose its practicality when we come to a level of technical systems. Any technical infrastructure shows this. What is the horizon of meaning for electricity? The system is so vast and is composed of innumerable industrial-machine systems, such as nuclear reactors and power lines, that the mystery of the complete system and its functioning is bound to leave a human in the dark. It brings electricity to our homes and for most of us that is enough, but it is not the truth about the system. The idea of a horizon of meanings is valuable as long as humans and machines are tied together in a situation in which the humans widen their powers through the tools or the machines they use; as is the case with an axe or an automatic hammer. But, once we move from centripetal systems towards centrifugal ones, the idea of a horizon of meanings loses much of its usefulness; humans and technological systems do not form a unit in a way that a human and a tool does. This does not mean that the systems would not be centripetal anymore; say made for humans. Certainly, they are exactly that. Life without them is unthinkable.

People may have only two attitudes towards these technological systems; either the attitude of a consumer or the attitude of a technician.

The consumer is interested in the products of the system, not of the system itself. The technician is interested in the system's functioning and its technical principles but not in the products it produces. Both of these, the functioning of the system and its end products, should be of concern if we would like to say this is what it means to have a real centripetal interest in the system. But, neither of those concerns deals with the functions of the technological systems outside the narrow horizon of consumption and maintenance. The true impact of technology on human life is left to mystery.

8. Machines Alone

The fourth category of technological environments, object centred and centrifugal, is probably the most typical one since the application of robotics to the industrial process took place. But still, it is rarely discussed, and when it is it's usually presented as a science fiction dystopia. The denial of this category is based on the false assumption that the machine cannot be conceived independently unless humans have some role to play in it. This notion is trivially true. The machine needs humans for its existence; just like there is no reason for speaking of language without someone speaking or having been speaking that language. Even still, there are some parts of a technical universe that the machines do better in by themselves without humans, just as there are syntactic rules for a language independent of its speaker's individual will (see Moravec, 1999).

Our example of a functional, object centred, and centrifugal technological environment is automation. Automation means the independence of the machine; its autonomy. In automation, the machine is by itself fulfilling its essence by doing what it is designed to do; just as Aristotle said humans should do to live a good and happy life. By following its own essence, the machine is autonomous and independent of others. The only way out of the horrors of Fordism and Taylorism is to replace humans on the assembly lines by robots; autonomous machines. When humans turn into machines in centripetal systems, it becomes clear that they can be replaced by machines as well. Factories are run by very few human workers who supervise the functioning of the machines through the machines, of course, instead of participating in the production process directly. Only when something goes wrong, God forbid, the situation turns from centrifugal into centripetal and humans are sent in to repair the machines and to evaluate the damage. This brings us to another advantage of automation. Machines can operate in environments where humans can not. The problem with centripetal environments is that the centre where humans are stuck is not necessarily a suitable place for humans. Besides that, there are situations that are, not only unhealthy, they are even deadly for ordinary humans. For example, the nuclear waste produced as a side effect from nuclear power generation needs to be stored somewhere, and whatever this place will be, it is certain that humans should not be there. These sorts of tasks should be left to machines. The most important thing for a machine is that it has some source of energy that it can recharge itself from. The rest is a question of building ever more adaptable machines to be able to operate in more complex environments than factories. At this moment, a machine can not even handle a simple drive on the surface of planet Mars. That is pretty pathetic. Engineers should get their act together. But, even more pathetic than that is the drive to send a man to the moon or to Mars; a sad reminder of the great human past because this is an operation which machines can now do better.

But, machines do a lot more than just replace humans in the tasks that require hard labour. As the machines' degree of autonomy increases, the machines start to communicate with each other, to repair each other, and in a dystopian view, finally go to war with each other. This is not metaphorical speech but the very basic idea of cybernetics (Hayles, 1999). The machine, say the disk drive in a computer, gives a certain output for the other machine, in this case the computers hard drive, which in turn gives its own output to the other machine, say the computer screen, and a picture appears to the users relief, unless, of course, there is a virus which interrupts the sent and received inputs and outputs. This machine's communication may not be human speech but it certainly resembles it. This kind of technological environment is clearly centrifugal. However, once we acknowledge the part that humans play in the machine's world, even the most object centred and centrifugal ones; we understand that human cannot completely be replaced by machines. This view, emphasising the human's role in these complex technological systems, still maintains that sometimes technologies are better left by themselves. This is called *technological realism*. The opposite view is *technological idealism*, which requires humans to constantly be in attendance with technology, if not as a physical person, then at least psychologically, imposing his values on all technologies. This is the traditional praxiological view.

For the technological idealist, the world is anthropomorphic, and the technology is completely dependent upon humans. But, although technology is trivially dependent on humans, it is the human endeavour that does not mean that it could not gain some degree of autonomy. The technological idealist argues that all speech about the autonomy of the machines should be interpreted metaphorically; when saying that machines operate alone in the places where human should not go, the situation is centrifugal only secondarily for those places are *a priori* places that humans should avoid.

9. Conclusion

The technological realist does not deny the fact that machines are dependent on humans. Perhaps they will be in the future, but not today. However, they acknowledge that despite this, both humans and machines have their own way of being in the world and the questions that we should be interested in have to do with the interrelationship between the machine-world and the human-world. The machines exist in their own horizon of meaning in which they can be classified in four categories, as we have attempted to show in this paper. The machine's autonomy corresponds with the human's autonomy without the former being reducible to the latter. The main problem with technological idealism is the prejudice that technology is merely instruments; something you may use or not use, whatever you wish.

This way of thinking about technology holds technology to be essentially *neutral*; that all the values technology bears are there because of the human intensions of its current user. But this instrumentalist way of looking into technology is severely limiting. It makes us blind far the changes that technology causes in us, its users, and our culture and values, that is, to technological determinism (Kaalikoski and Airaksinen, 1994). Only by accepting the fact that machines have a way of being of their own can we become more alert of the problems that their use and presence poses on us. To make it more elementary, if we were stuck to the neutral picture of technology in the case of automatism, for example, we probably would not have arrived at the situation where robots are left by themselves in the factories while the human workers are performing other kinds of jobs, say, in the services. For robot technology to be present in the factories requires Taylorism as its starting point. Only after humans have developed a "science of management," tried it on ourselves and found it unsatisfactory, will the machines take over that job. In the case of neutral technology we are bound to the situation where all ethically sound technologies are merely extensions of our human organs, enabling us to perform our routine tasks with more efficiency. All other kinds of technologies would be judged ethically untenable as they are designed for jobs that humans never would or should do. Thus, only by acknowledging some degree of autonomy for the machines, which means to acknowledge their own ways of acting, can we arrive at our present situation with all the problems it contains but also, with all the potential it offers. And because we have already arrived at the situation where technology has gained a degree of autonomy alongside with humans and there is no going back without a drastic decrease in our standard of living, we should acknowledge the facts and start thinking on the basis of the facts towards a new ethics of technology.

Notes

1. This paper is loosely based on some ideas in Timo Airaksinen's book *The Future of Human Machine* (WSOY, 2006, in Finnish). The authors thank TEKES for a generous research grant and Mr Roy Holcombe for his expert help.

References

Bauer, M.E. 1997. *Resistance to New Technology.* Cambridge University Press, Cambridge.
Brooks, R. 2002. *Robot.* Allen Lane Penguin Press, Harmondsworth.
Clark, A. 2003. *Natural Born Cyborgs.* Oxford University Press, New York.
Gadamer, H.G 1975. *Truth and Method.* Continuum, London.
Gigerenzer, G. 2002. *Reckoning with Risk.* Penguin Books, Harmondsworth.
Gramsci, A. 1971. *Selections from the Prison Notebooks.* Eds. Quintin Hoare and Geoffrey Nowell Smith. Lawrence and Wishart, London.
Heidegger, M. 1977. *The Question concerning Technology and Other Essays.* Harper Torchbooks, New York.
Hayles, N.C. 1999. *How We Became Posthuman.* Chicago University Press, Chicago.
McCarthy, J. and Wright, P. 2004. *Technology as Experience.* MIT Press, Cambridge Ma.
Kaalikoski, K. and Airaksinen, T. 1994. „Instrumental Rationality", *Proto Soziologie* no 6, pp. 151–159.
McLuhan, M. 1964. *Understanding Media: The Extensions of Man.* Signet book, New York.
Moravec, H. 1999. *Robot.* Oxford University Press, New York.
Oudshoorn, N. and Pinch, T.E. 2003. *How Users Matter: The Co-Construction of Users and Technology.* MIT Press, Cambridge Ma.
Pacey, A. 1999. *Meaning in Technology.* MIT Press, Cambridge Ma.
Parenti, C. 2003. *The Soft Cage: Surveillance in America.* Basic Books, New York.

Material Culture and the Conditions for the Good Life: Albert Borgmann on Thinking about Ethics in Technological Culture

Topi Heikkerö
Centre for Social Ethics
Department of Systematic Theology
University of Helsinki
Finland

We shape our buildings;
thereafter they shape us.
Winston Churchill

Devices, that was the claim,
dissolve the coherent and engaging character
of the pretechnological world of things.[1]
Albert Borgmann

1. Introduction

In what sense does more advanced technology provide a better life? The question is awkward: we are not used to thinking about technological progress and the good of life in the same sentence. Our culture in general appears to take new technologies for granted. Novel innovations are in no need of legitimation on the grounds that they improve the quality of life, or we perceive as self-evident that better technology means better life. Technology policy specialists most often treat this question within the framework of cost-benefit-risk analysis. Its quantitative approach, however, fails to truly recognize such a qualitative issue as reflection on the good life. How do academic viewpoints relate to this question? In academia, there exists a newly-born discipline of science, technology, and society (STS) studies. This discipline has created a number of methods for studying technology as a socially embedded phenomenon. But since the common starting point of these approaches is scientific disengagement, or "value-neutrality," they tend to evade the question posed above.[2] Within philosophy, finally, we find some discussions on the topic of technology and the good life, albeit few of them – and even these few take place far from the limelight of the academic community.[3]

In this essay, I address the topic through an introduction to Albert Borgmann's thought. Borgmann is a North American (native German) philosopher who has analyzed technology as a social, ethical, and political phenomenon within the past 30 years. Borgmann's thinking is rooted in the phenomenological tradition of contemporary philosophy. While his analysis clearly draws on Martin Heidegger's philosophy of technology,[4] Borgmann has created a unique position that makes use of various other ingredients. He argues that in contemporary culture, genuine ethical thinking is possible only in conjunction with an analysis of the technological scheme prevalent in production and shaping the way of life.

I will first set the question in a slightly wider historical perspective, after which I will introduce Borgmann's philosophical tools for treating the question. I will then continue to the possibilities for ethical reflection in contemporary technologized culture and Borgmann's remedies for what he perceives as the malaises of this condition. Finally, I will conclude with some critical remarks about the relevance and problems of Borgmann's project.

2. Does Better Technology Mean Better Life?[5]

In the beginning of the Modern Era, a new perspective emerged in science and philosophy. Thinkers who participated in the formation of the novel natural scientific framework explicitly held that this pragmatically-oriented technoscience[6] would provide humanity a better life. In *Discourse on the Method*, René Descartes articulates this ethos as follows:

> Through this philosophy we could know the power and action of fire, water, air, the stars, the heavens and all the other bodies in our environment as distinctly as

we know the various crafts of our artisans; and we could use this knowledge – as artisans use theirs – for all the purposes for which it is appropriate, and thus make ourselves, as it were, the lords and masters of nature. This is desirable not only for the invention of innumerable devices which would facilitate our enjoyment of the fruits of the earth and all the goods we find there, but also, and most importantly, for the maintenance of health, which undoubtedly is the chief good and the foundation of all other goods in this life.[7]

Moreover, during the Enlightenment period thinkers linked technoscientific progress to political progress, manifest in expansion of liberty, equality, and democracy wider in society.[8]

The promise Descartes and others presented has largely come to pass: technoscientific progress has enabled the wide-scale spread of democracy, well-being, and equality – in developed societies. Moreover, this same advance appears to have granted us greater freedom in terms of self-realization. This development, however, has created serious problems as well. The ecological crisis that has followed technological mass production is, arguably, the most severe manifestation of this predicament. Our focus here, nonetheless, is on the conditions of the good life.[9] Although at first glance the promise of technology, with respect to human life, appears to have proven true and beneficial, upon closer examination the situation is more ambivalent.

This ambivalence is precisely the problem Borgmann addresses in his work, for he points out a number of phenomena in which this ambivalence is manifest. For instance, although institutional science progresses, people's scientific world view is not actually improving. Despite the increase in popular access to nutritional knowledge and in opportunities for exercise, obesity has become an ever-growing problem and the physical condition of the population is degrading. The more developed a society becomes, the more divorce rates tend to rise. The public awareness of communal, national, and global issues appears to remain stagnant despite economic growth, technological innovations in communication devices, and more leisure. In Borgmann's reading, the corpus of social scientific research data documents these trends, including some studies which indicate that citizens of the developed world profess to be no happier than those in developing countries.[10]

Borgmann does not hesitate to pass a value judgment on the situation. He finds the state of contemporary culture depressive. Borgmann, however, intends not to compare our contemporary culture to that of some earlier epoch, but to evaluate our culture's achievements against what appear to be its own inherent ideals and possibilities.[11] Thus, his analysis suggests that measured both by historically documented ideals and people's inner convictions contemporary culture fails to achieve the good life. Borgmann argues that if one looks at the members of contemporary culture in the light of such traditional western ideals as knowledge of reality, physical and mental excellence, citizenship, familiarity with the arts, or charity, it fails – especially if the assessment takes into account its apparent possibilities. This same failure arises in studies of people's subjective perceptions. For instance, research on TV watching indicates that despite the increase (over the years) in time spent in front of the television set TV-watchers have found this pastime unrewarding and in no way excellent. In fact, the term "couch potato" reflects this judgment.[12] Thus, it appears that people's actual ways of living and their deepest convictions exist in disharmony. In short, though our culture seeks the good life through technological innovation, this pursuit often leads to frustration and an undefinable emptiness. Borgmann sheds light on this phenomenon through his analysis of technological commodification.[13]

3. Material Culture in the Technological Era

Borgmann defines material culture as "the totality of tangible artifacts that humans have produced."[14] The character of material culture has evolved as a result of technological revolutions. Although technological devices and systems vary both in historical succession and in synchronic perspectives, Borgmann argues that one can nevertheless distinguish "the pattern of technology." This pattern, which he names *the device paradigm,* is the general way of taking up with reality in technologized society. The pretechnological culture[15] took up with the world in a different manner; the device paradigm is the modern, technologized way to be in the world.[16]

Borgmann's telling example concerns warmth. In pre-modern days, warmth was produced by a hearth, which kept the temperature in the house comfortable and provided heat for baking and cooking. An old-fashioned stove, however, required care as well, for someone needed to select, saw, and split the wood, carry it in, and keep an eye on the fire in order to maintain a suitable temperature. All this required skill and fostered a small social network centered around the stove: numerous tasks had to be divided within the family. The hearth was also a particular thing, a metal or stone construction usually placed in the center of the house. It served as a place where the family gathered for warmth, food, and often just to chat. While this may sound romantic, taking care of the hearth was burdensome, and sometimes a real hassle. Wood could be in short supply, the wood might be green or wet, quarrels could arise over the sharing of tasks, and all this in addition to the heavy workload.[17]

Today, our houses no longer have hearths. We have been disburdened from the tasks of heating our homes. In most cases, this has occurred gradually, culminating in the installation of a central heating system. Normally, we seldom even know where the heating plant is located; we just receive the commodity it produces – warmth with neither the hassle, the smell of smoke, the concrete object, and nor the social relations. Such is an actualization of the device paradigm.[18]

The following aspects characterize the device paradigm: disburdenment, disengagement, de-contextualization, the availability and disposability of commodities within it, and discontinuity.[19] Technology disburdens us by providing us commodities without effort or skill, any time, and anywhere. If one wanted to enjoy Mozart's *Don Giovanni* in Copenhagen in the 1840s, one had to find a theater where the piece was played, wear suitable clothing, buy a ticket, be present in the right place at the right time, and interact in the social context of the opera. In our contemporary situation, we have witnessed innumerable phases of development in analog and digital audio technology as well as advances in video production and distribution from motion pictures to high definition television. Thus, a contemporary Dane can enjoy *Don Giovanni,* beginning with a favorite aria, with CD or DVD equipment and a surround sound system.[20] Technology disengages and de-contextualizes the desired commodity – whether warmth or music – from all other aspects.

Borgmann refers to discontinuity as the tendency of the device paradigm to create sharper polarities than pre-modern ways of encountering reality. These polarities include means/ends, body/mind, interior/exterior, work/leisure, and individual/communal. In short, the supposition is that these phenomena were more intertwined in the pretechnological ways of being in the world, and that the device paradigm abstracts them from each other. Listening to a *Don Giovanni* CD is an end in itself, a mental, interior, and individual experience, but a night in the opera is an organic whole that transcends such categories.

The disengaging character of the device paradigm sums up Borgmann's argument. Because devices fail to engage the human beings who use them, technological disburdenment undermines the depth, integrity, and meaning of life.

4. Ethical Thinking and Material Culture

Based on his analysis of technology, Borgmann points out shortcomings in contemporary ethical and political thinking. Concepts such as "value" and "self-realization" are vague, if not empty, unless the technological pattern underlying them is made visible. Borgmann argues that when political liberalism avoids defining the good life and instead leaves this task to individual reflection and choice, the device paradigm silently, without discussion or conscious reflection, defines the good life. Self-realization is then reduced to a mere consumer choice. Value is measured through bargaining with ever-present techno-economic realities or it equals with arbitrary subjectivity in their absence.[21]

The roots of this problem are manifold and complex. One root is the classical notion of morally neutral production. Borgmann considers the Aristotelian distinction between doing (*praxis*), which is an end in itself and making (*poiesis*), which carries more a productive or instrumental purpose,[22] untenable in the modern context. *Praxis,* or practical doing, is understood to be morally-laden, while *poiesis,* or productive making, is morally neutral. This distinction creates some of the most severe problems and misunderstandings in the philosophy of technology:[23]

> Some observers of modern culture do note disparities and tensions between contemporary doing and making, between ethics and technology. They complain that, while technology has advanced dramatically, our moral aptitude in dealing with it has not. But this complaint is as radically mistaken as the general divorce of doing from making. It fails to see that a technological accomplishment, *the development and adoption of a technological device always and already constitutes a moral decision.*[24]

Why should *poiesis* be rethought as morally-laden? The simplest answer is that design, development, and production of technological devices always aim to produce some goods and is therefore value-laden. Technologies, moreover, are more than mere usable tools; they are, as Borgmann's analysis of device paradigm pointed out, most appropriately characterized as systems that have an effect on human users, their thinking, skills, and engagement in reality.

With regard to contemporary thinking, however, one look at the academic community reveals that ethical reflection and awareness of technological realities rarely coexist. First, applied ethics, mostly practiced within analytic tradition, appears to be somewhat blind to technology as an extensive and existentially conditioning phenomenon. Second, as mentioned in the beginning, technology studies that currently create nuanced analyses of their subject matter operate within a framework that most often excludes normativity. Third, continentally-oriented philosophy, no doubt, is sensitive to phenomena that determine the human condition, but often lacks the perspective of ethical agency present in analytic ethics, and professes little or no empirical approach to technology.[25]

To sum up, Borgmann's argument concerning ethics and technology suggests that traditional ethical thinking, which assumes that singular individuals make rational decisions, misses the manner in which material culture shapes ethical agency. Therefore, besides considering the individual, ethics should include reflections about the places, organizations, and commodities that form the human-made world in which ethical subjects live.

5. Ideas for a Reform

As previously noted, Borgmann argues that the character of contemporary life lacks focus, integrity, and excellence, and that the distinct cause of these problems lies in the disengagement inherent in the device paradigm. In short, he sees the device paradigm as tragically misplaced if it is to serve as a tool for acquiring the good life. Yet, according to Borgmann, the device paradigm has its place in the overall infrastructure.[26]

On the practical level of everyday existence, Borgmann's tools for regaining focus, integrity, and excellence are focal things and practices. Consider the following examples. The wheelwright's craft one hundred years ago had to do with focal things, cart wheels, whereas the manufacture of car wheels on an assembly line represents more of an actualization of the device paradigm. Fly-fishing can be a focal practice, then a fishing rod and/or trout is the focal thing. For a runner, the running route is the focal thing, for a backpacker, the wilderness. As Henry David Thoreau described in his essay "Walking," the simple act of walking can be a focal practice. Further, arts and crafts can be focal practices.[27]

A family dinner is also a focal practice, and all the material objects it involves are focal things within that practice; fast food and ready-made meals are its equivalents within the device paradigm. Borgmann, in fact, discusses "the culture of the table," which can include much more than just cooking and eating: it has a beginning and an end, it forms a tangible event for which the family congregates. This "culture of the table" blends with traditions of cuisine, and in religious families involves grace through which family members express their gratitude for the gifts of the earth.[28]

Consider now focal things and practices from an analytical perspective. Borgmann remarks that while dealing with focal things one is guided by a commanding presence that exactly counters the distraction produced by devices. Within the device paradigm everything is controlled; everything is made predictable. In contrast, otherness lives within focal practices. Focal practices require exertion, skill, self-transcendence, perseverance, endurance, patience, commitment, and attention – qualities from which the device disburdens us. In this manner, Borgmann suggests that focal practices can recenter our lives and provide contexts where virtue and excellence can be worked for. Discontinuity characterizes activities within the device paradigm whereas focal practices lead to unification and integration: in fly-fishing, body and mind work together; the runner merges in the environment in which he or she is running; for the hiker, in backpacking, culture and nature – as well as the hiking traditions and the present – converge. Means and ends merge as well; in most focal practices they cannot be strictly separated. Running is simultaneously an end in itself and a means to maintain well-being, as Borgmann describes: "In the runner, effort and joy are one; the split between means and ends, labor and leisure is healed."[29]

Moreover, the role of focal things in Borgmann's theory is not only to supercede the device paradigm in providing a better sphere of living, but to enable the proper evaluation of technology:

> A focal practice engenders an intelligent and selective attitude toward technology. It leads to the simplification and perfection of technology in the background of one's focal concern and to a discerning use of technological products at the center of one's practice.[30]

Borgmann's reflection on technology works within the contrast of the device paradigm and the focal reality. Reform based on focal things and practices is more than a mere transformation in thinking: Borgmann explains us not how to think but how to engage in reality. At the same time, his anthropology stays fairly open, as the good life in his sense refers to the process of interaction with things; it pursues no distinct ideological goal. Borgmann's political ideas are based on the conclusion that only within the contexts of focal things and practices is the good life possible.[31] Therefore, technology policy that takes the quality of life into account, and not just quantitative standard of living, should create societal arrangements that enable citizens to engage as much as possible in focal interaction. In political thinking, this requires that the moral dimension of material culture be acknowledged alongside the analysis of power relations and economic phenomena. In ethics, Borgmann argues for the deeper consideration of practices and ways of life in addition to individual, instantaneous moral decisions.

6. Critique and Reflection

Borgmann's approach raises three critical considerations. First, Borgmann's ability to capture the pattern, paradigm of modern technology is remarkable, particularly given the countless number and diversity in character of technological devices and systems. Nevertheless, Borgmann appears to underestimate the variety of technological phenomena. Andrew Feenberg and Larry Hickman have criticized Borgmann for ignoring the results of technology studies. Moreover, Borgmann's view of the human use of devices in light of one paradigm fails to do justice to the varieties of consumer reception. Numerous case studies indicate that consumers are often creative in adopting new technologies and are quick to find unintended and subversive ways to employ them. The Internet, for instance, has become a platform for democratizing processes as well as a tool for marginal groups across the globe.[32] Furthermore, one cannot help asking how sharp is the difference in engagement between things and devices. On the one hand, is it impossible to have an engaging relationship with one's CD player or iPod?[33] On the other hand, the culture of the table can serve as a stage for artificiality, disengagement, or being tuned-out.

These problems raise a second question: What exactly is a focal thing or a focal practice? How can we discern a focal thing from a mere device? How can we know if things have integrating power?[34] Larry Hickman has proposed that focal practices and their wholesome effects should be empirically tested.[35] Borgmann has rejected this possibility. According to him, we may incorrectly evaluate focal things and practices, but in principle they defy objective testing.[36] Borgmann's position is understandable, since testing focal things would entail some kind of operationalization, or even quantification, which would contradict the highly context-related nature of focal things and their limitless variety of aspects. The inability to evaluate focal things without ambiguities, however, means that questions concerning them remain open.

Focal practices are also susceptible to another general criticism. An uncharitable reader could ask: "Is he suggesting that 'good hobbies' will save us from the problems of technological modernity?" In fact, Borgmann does have substantial ideas regarding larger socio-economic settings that relate the role of focal things in our way of life.[37] Consequently, the uncharitable question largely misses the point. Nevertheless, if one considers, for instance, farming a focal practice as opposed to running or fly-fishing, farming is evidently a total way of life and livelihood, while the others are simply hobbies.[38] The problem is that farming as a livelihood is no longer a reality for – and is probably not even desired by – most people.[39] The same goes for other traditional occupations, such as those of an artisan or even a homemaker.

My third point is a partial answer to the questions related to the second point. A cultural comparison could offer a solution to the above-mentioned worries.[40] Traditional Japanese culture, for instance, espouses various practices possessing explicit, conscious focal character: *Kendo,* the Way of the Sword, and *Chado,* the Way of Tea, present the focal thing – the focal practice structure even in their names. The aim of cherishing the focal things within focal practices is also evident in the art of gardening, flower arrangement, calligraphy, and brush painting. The main thrust of these practices – "dos" or ways – is to come to know oneself, others, and reality more deeply through practicing the respective art. The intention behind Borgmann's concepts is exactly the same. In Zen Buddhism, even quotidian mundane tasks, such as cooking, chopping wood, and carrying water, are viewed as spiritual practices.[41] Thus, for a Zen practioner, disburdenment from these mundane tasks would result in a lost opportunity of spiritual growth. The close resemblance between these Japanese concepts and Borgmann's concerns suggests that Borgmann's ideas are not as idiosyncratic as they may at first seem.

Of course, unreflected focal things and practices abound in all cultures. Consider, for instance, sports, sauna culture, arts, story-telling, pilgrimages and other religious activities. In Japan, however, a conscious, reflected philosophy of focal things and practices exists, which is exactly what Borgmann perceives necessary in the cultural condition in which we have become disillusioned of the good life providing powers of technology.[42]

7. Concluding Remarks

Let us now revisit our initial question: Does more advanced technology provide a better life? If democracy, equality, and freedom from sickness, poverty, and toil constitute a better life as a whole, the answer is a resounding "Yes." This essay, however, has focused on the less visible features of the technological betterment of life, both its hidden dangers and its inherent possibilities. In conclusion, technological disburdenment indirectly provides the potential to attain the good life, but to see this disburdenment itself as the way to a better life instead leads one in the opposite direction, toward a shallow existence and one-dimensional culture.

In academic disciplines, such perceptions could lead toward a wider conception of ethics, toward an ethical thinking that is better aware of the institutional and material arrangements that condition moral agency. Moreover, keeping the question of the good life open and removed from philosophical discussions appears untenable. If we do not explicitly discuss the question, we submit to the silent answer granted by technological paradigm. In this context, Borgmann's way of addressing the question of technology's relationship with the good life in his concrete and non-paternalistic manner reveals its relevance.

Acknowledgments

I thank Albert Borgmann for generously providing time to discuss his work. I also thank Finnish Cultural Foundation for its grant which enabled me to concentrate on writing this article. I thank Stephen Stalter for revising this text. Further, I am grateful to Carl Mitcham, David Strong, Jaana Hallamaa, Sami Pihlström, Timo Airaksinen, Janne Kylliäinen, Wha-Chul Son, Karin Christiansen, Tomi Kiilakoski, Markku Valtanen, and Scot Yoder for their suggestions, discussions, and correspondence, all of which contributed to this essay. An earlier version of this paper was presented, under the same title, at an annual meeting of *Societas Ethica*, in Salzburg, Austria, in August 2005. Any faults are solely mine.

Notes

1. Borgmann 1984, 47.
2. For a critique of value-neutrality in STS, see Winner 1993.
3. For a history of philosophy of technology, Carl Mitcham's (1994) *Thinking through Technology* is the most informative volume I have found. For the current state of this subdiscipline, see Ihde 2004.
4. Dreyfus (1993) maps out in a concise manner Heidegger's analyses of nihilism and technology, which closely parallel the themes of this essay.

5. Cf. Marx (1987), "Does Improved Technology Mean Progress?" See also Mitcham (1994, 275–299) "Three Ways of Being with Technology" and Thomas Hughes's *Human-Built World* (2004) for historical overviews of attitudes to technology.
6. Here I start from an assumption that science and technology cannot be historically or socially separated from each other. This, however, does not suggest the impossibility of such a conceptual distinction. The term "technoscience" refers to coexistence and co-development of science and technology. This position is obviously debatable. See, for instance, Mitcham 1994, 199–204; Borgmann 1984, 17, 29–31, 50; Hickman 2001, 3,5, 41–43.
7. Descartes 1985, 142–143. Other pertinent formulations of this idea appear in the writings of Francis Bacon (1960) and Thomaso Campanella.
8. Marx 1987. Hughes 2004, 17–43. For instance, in 1813, Thomas Jefferson wrote to John Adams on the effects of the new science and the American Revolution in Europe: "Science had liberated the ideas of those who read and reflect, and the American example had kindled feelings of right in the people. An insurrection has consequently begun, of science, talents, and courage, against rank and birth, which have fallen into contempt... Science is progressive." Quote from Marx 1987.
9. Ecological problems and the question of the good life are related in many complex ways. I can, however, point out two obvious connections: First, both are creations of modern technoscientific advancements, and second, technological disengagement in personal lifeworld often results in a person's lack in awareness of ecological realities.
10. Borgmann 1984, 124–143.
11. Borgmann 1984, 124–143.
12. Borgmann 1984, 140–143.
13. Borgmann 1984, 38–39, 172–3; 1992, 110–116; 1995; 2003, 11–34; 2005.
14. Borgmann 2005.
15. Borgmann's analysis of technology obviously addresses modern technology, not all historical instances of human techniques.
16. Borgmann 1984, 40–48, 53, 76–77, 146–48; 2003, 17–18, 31, 121–22. Strong & Higgs 2000, 27–37. Borgmann's device paradigm, no doubt, echoes Heidegger's (1977, 19 ff) Enframing *(Ge-stell)*. Additional resemblances to Jacques Ellul's *La Tecnique* and Lewis Mumford's monotechnics may also exist. Borgmann (1984, 7–15), however, has always sought to distance himself from these "technophobic" thinkers and their substantive understanding of technology. His critics believe he has failed to distance himself sufficiently (see later in the text body).
17. Borgmann 1984, 41–42.
18. Borgmann 1984, 42–43.
19. Borgmann 1984, 40–48, 76–77; 1992, 82–97; 2003, 16–19.
20. For reflections on the compact disc technology see Borgmann 1999, 20, 195, 120, 181–182.
21. Borgmann 1984, 89, 213–4; 1992, 40–45. Taylor 1992, 55–69. Winner 1986, 155–163. Power 2000.
22. *EN* 1139a26–b9; *Met* 1025b18–26, 1064a10–18.
23. Borgmann 1992, 110 ff; Strong 2000, 321–322.
24. Borgmann 1992, 110; emphasis mine.
25. Mitcham & Nissenbaum 1998. Keulartz et al. 2004, 3–14. Winner 1993.
26. Borgmann 1984, 226–230, 237 ff.
27. Borgmann 1984, 44–47, 201–204. Maclean 1976. Thoreau 1964, 592–630.
28. Borgmann 1984, 204–206. Morgan Spurlock's tongue-in-cheek documentary *Super-size Me* (2004) examines the fast food culture in a drastic light. Although

Spurlock concentrates on the medical side of the phenomenon, he nevertheless shows cultural background and consequences of fast-food culture. In contrast, Gabriel Axel's *Babette's Feast* (1987) is one of the best cinematic elaborations on the culture of the table. *Slow Food* is an organization that promotes the protection of the "pleasures of the table from the homogenization of modern fast food and life." It was founded in 1986 in Italy and now boasts 80 000 members in over 100 countries. It promotes education, gastronomic culture, agricultural biodiversity, and the protection of traditional foods. Their web site: http://www.slowfood.com/

29. Borgmann 1984, 202. Borgmann's debts to Heidegger are many. For instance, one can see as a goal of focal practices Heideggerian *Eigentlichkeit (Being and Time,* Chapters 51-63). Second, in his contention that the device paradigm is not "bad" in itself but instead requires a proper role in our culture, Borgmann appears to give concrete meaning to Heidegger's (1977, 26–35) view that demonizing technology is useless. Third, Borgmann's thinking on things derives from Heidegger's (2001, 161–184) poetic elaborations on *Dinge,* in which the fourfold – earth, sky, mortals, and divinities – gathers.
30. Borgmann 1984, 221.
31. Borgmann's idea of the good life has critical potential without being overly essentialist. He suggests that there exists a plurality of focal things, concerns, and practices (1984, 186, 211–13, 219).
32. Feenberg 2000, 298–305, 308–311. Hickman 2000, 93 ff; 2001, 115–119.
33. In his cult novel *Zen and the Art of Motorcycle Maintenance* (1974), Robert Pirsig claims that working with a motorbike and other devices can be a vehicle for attaining self-knowledge and peace of mind.
34. Brittan 2000.
35. Hickman 2000, 98–99.
36. Borgmann 2000, 356–357.
37. Borgmann 1984, 157–169, 226–246.
38. Thompson 2000, 167–178.
39. If organic farming and less technologically-enhanced forms of agriculture gain wide popularity, slightly more labor-intensive ways of farming could see a re-emergence. Nevertheless, the current trend in the transformation of the occupational structure toward services and information processing will likely remain unaffected.
40. Dreyfus's (1993, 295, 307) remarks on Japanese culture are similar to my observations.
41. Dogen 1985, 53–68, 124–128. Suzuki 1956, 279ff.
42. Heikkerö 2005.

References

Aristotle. 1947. *Nicomachean Ethics.* Transl. H. Rackham. Cambridge, MA: Harvard University Press.

Aristotle. 1948. *Metaphysics.* Commentary and Introduction W.D. Ross. Oxford: Oxford University Press.

Bacon, F. 1960. *The New Organon and Related Writings.* Ed. Fulton H. Anderson. New York: The Liberal Arts Press.

Borgmann, A. 1984. *Technology and the Character of Contemporary Life: A Philosophical Inquiry.* Chicago: The University of Chicago Press.

Borgmann, A. 1992. *Crossing the Postmodern Divide.* Chicago: University of Chicago Press.
Borgmann, A. 1995. "The Moral Significance of the Material Culture" in Feenberg and Hannay 1995.
Borgmann, A. 1999. *Holding On to Reality: the Nature of Information at the Turn of the Millennium.* Chicago: University of Chicago Press.
Borgmann, A. 2000. "Reply to My Critics." *Technology and Good Life?*, 341–370.
Borgmann, A. 2003. *Power Failure: Christianity in the Culture of Technology.* Grand Rapids: Brazos Press.
Borgmann, A. 2005. "Material Culture" entry in *The Encyclopedia of Science, Technology, and Ethics.* Detroit: Macmillan Reference.
Brittan, G. 2000. "Technology and Nostalgia" in Higgs et al. 2000, 70–88.
Cutcliffe, S.H. & Mitcham, Carl (ed.) 2001. *Visions of STS: Counterpoints in Science, Technology, and Society Studies.* Albany, NY: State University of New York Press.
Descartes, R. 1985. *The Philosophical Writings of Descartes, Vol. I.* Transl. John Cottingham et al. Cambridge: Cambridge University Press.
Dogen, E. 1985. *Moon in a Dewdrop: Writings of Zen Master Dogen.* Transl. Kazuaki Tanahashi. Canada: HarperCollins Canada.
Dreyfus, H.L. 1993. "Heidegger on the connection between nihilism, art, technology, and politics" in Guignon 1993, 289–316.
Feenberg, A. 2000. "From Essentialism to Constructivism: Philosophy of Technology at the Crossroads" in Higgs et al. 2000, 294–315.
Feenberg, A. & Hannay, A. (ed.) 1995. *Technology and the Politics of Knowledge.* Bloomington and Indianapolis: Indiana University Press.
Guignon, Ch.B. (ed.) 1993. *The Cambridge Companion to Heidegger.* Cambridge: Cambridge University Press.
Heidegger, M. 1977. *The Question Concerning Technology and Other Essays.* Transl. William Lovitt. New York: Harper & Row.
Heidegger, M. 1978. *Sein und Zeit.* Tübingen: Max Niemayer Verlag.
Heidegger, M. 1996. *Being and Time: A Translation of Sein and Zeit.* Transl. Joan Stambaugh. Albany: State University of New York Press.
Heidegger, M. 2001. *Poetry, Language, Thought.* Transl. Albert Hofstadter. New York: HarperCollins.
Heikkerö, T. 2005. "The Good Life in a Technological World: Focal Things and Practices in the West and in Japan" in *Technology in Society*, Vol. 27, No 2, 251–259.
Hickman, L. 2000. "Focaltechnics, Pragmatechnics, and the Reform of Technology" in Higgs et al. 2000, 89–105.
Hickman, L. 2001. *Philosophical Tools for Technological Culture: Putting Pragmatism to Work.* Bloomington: Indiana University Press.
Higgs, E., Light, A. & Strong, D. (ed.) 2000. *Technology and the Good Life?* Chicago & London: University of Chicago Press.
Hughes, T. 2004. *Human-Built World: How to Think about Technology and Culture.* Chicago & London: University of Chicago Press.
Ihde, D. 2004. "Has the Philosophy of Technology Arrived? A State-of-the-Art Review" in *Philosophy of Science,* Vol. 71, Number 1 (2004), 117–131.
Keulartz, J., Schermer, M., Korthals, M. & Swierstra. 2004. "Ethics in Technological Culture: A Programmatic Proposal for a Pragmatist Approach" in *Science, Technology, & Human Values,* Vol. 29, No. 1, 3–29.
Marx, L. 1987. "Does Improved Technology Mean Progress?" in *Technology Review,* January 1987, 33–41.

Mitcham, C. 1994. *Thinking through Technology: the Path between Engineering and Philosophy.* Chicago: University of Chicago Press.

Mitcham, C., & Nissenbaum, H. 1998. "Technology and Ethics" in *Routledge Encyclopedia of Philosophy.* Vol. 9.

Pirsig, R.M. 1974. *Zen and the Art of Motorcycle Maintenance: An Inquiry into Values.* New York: William Morrow.

Power, T.M. 2000. "Trapped in Consumption: Modern Social Structure and Entrenchment of the Device" in Higgs et al. 2000, 271–293.

Strong, D. 2000. "Philosophy in the Service of Things" in Higgs et al. 2000, 316–338.

Strong, D. & Higgs, E. 2000. "Albert Borgmann's Philosophy of Technology" in Higgs et al. 2000, 19–37.

Suzuki, D.T. 1956. *Zen Buddhism: Selected Writings of D.T. Suzuki.* New York: Doubleday.

Thompson, P.T. 2000. "Farming as Focal Practice" in Higgs et al. 2000, 166–181.

Thoreau, H.D. 1964. *The Portable Thoreau.* Ed. Carl Bode. New York: The Viking Press.

Winner, L. 1986. *The Whale and the Reactor: A Search for Limits in an Age of High Technology.* Chicago and London: The University of Chicago Press.

Winner, L. 1993. "Upon Opening the Black Box and Finding it Empty: Social Constructivism and the Philosophy of Technology" in *Science, Technology, and Human Values*, Vol. 18, No. 3, 262–378.

Technology Changing Medical Ethics[1]

Tuija Takala
University of Manchester, U.K.
University of Helsinki, Finland

1. Introduction

Vast advances in medical technology are challenging traditional medical ethics. The concepts of health and disease are underminesd by improved methods of diagnosis and treatment, and with the rapid advances in genetics, they might lose their old meanings altogether. Ever since the processes of *in vitro fertilisation* were introduced, old ideas concerning the beginning of human life have been challenged. With the now looming possibilities of human cloning, the effects might be even more profound. Developments in life-sustaining technology have for long shifted back the boundaries of death, and many are looking forward to future possibilities of expanding the human life span considerably further. Issues of chance have become issues of choice, and the choices are getting more complicated.In this chapter the profound impact of technological advances on medical practice and its ethics are studied. It is shown how the ethics of medical practice is gradually shaped by the technology available. Technological advances challenge the profound concepts, and the ethical paradigm is reshaped by the new definitions. Technological advances have profound impacts in all fields of medicine. In this article I will use as examples the changes that technology has made to the ethical questions encountered in beginning and end of life situations. In the latter part of this paper I will take a critical look at the paradigm shift that the new biotechnologies are bringing about in attitudes towards consent and individual rights, and argue that we should proceed with caution.

2. Medical Ethics in Historical Context

If medical ethics is understood as the proper code of conduct of medical professionals, it has been around as long as the profession itself has existed. The profession's ancestors, the shamans, witches, herbalist and others are believed to have followed their own sets of values even before the times of written language. The earliest written example known to us of medical ethics can be found in the law of Hammurabi, which, as far as we know, dates back to 1800 BCE Babylon. Of the ancient codes, the c. 500 BCE pledge often referred to as the *Hippocratic Oath* still partly guides modern medical practice. There are, of course, paragraphs in the oath, such as the promise to honour one's teacher like a parent and, if need be, supply him with the necessities of life, that are seldom upheld in modern settings.[2]

Throughout the centuries that Christianity in its various forms ruled the Western World, it shaped the ethics of the medical profession. But also the professionals themselves had their own guilds and associations, and their own codes of conduct. By the beginning of the 20th century most countries had established medical associations of their own. It was, however, the atrocities of the Nazi doctors during World War II that gave rise to the umbrella organisation of all medical associations, the *World Medical Association*. At this point it was thought necessary for the medical profession to unite, to distance themselves from the deeds committed by the Nazi doctors and to establish an oath or a pledge of ethical conduct to be taken by all those who practice medicine. The first ethical code agreed by the World Medical Association became known as the Declaration of Geneva. Other codes of conduct have further been specified in the Declarations of Helsinki and Tokyo.[3]

Alongside the codes of the medical doctors, the codes of nurses have also had an impact on the ethical codes in use at health care settings. In the 19th century Florence Nightingale established the first nursing school in England and in 1899 the *International Council of Nurses* was established. The ethics of care has always been one of the key features of associations of nurses.[4]

It was not, however, until the late 1960s that medical ethics became a subject of wider public discussion. The early Declarations of the World Medical Association had been drafted by the medical professionals themselves and the professionals were trusted to be the guardians of their own codes of conduct. During the last decades of the 20th century this was no longer considered to appropriate. The industrialised world saw the rise of a new generation of individuals who refused to be subjected to medical paternalism – they had values of their own. The medical profession and its ethics were attacked both from the liberal and conservative camps. People felt that it should not be up to doctors to decide when abortion was justified or when treatment became futile. The questions of abortion and euthanasia became topics of heated public debates all around the Western world, and bioethics, as a field of academic study, was born.[5]

The past 35 years have seen earth-shattering technological advances and the future promises even more. Especially in the field of medicine, technology has a profound and direct impact on the very essentials of life, and causes great ethical confusion.

3. Beginning of Life

The moral status of the embryo is a question that has been debated throughout the centuries and on which no general agreement seems to be forthcoming. Some argue that human life begins at conception and should be protected from that moment. Others have argued that it is when the fertilised egg reaches the blastocyst stage (this is when the cells of the embryo have just started to differentiate) some five to seven days after fertilisation that marks the point. Some say that a human embryo is entitled to some protection, but not equal to beings with human genomes who are already outside their mothers' bodies. The list of the various normative positions on the moral status of the embryo is almost endless.[6]

For centuries, the discussion on the moral status of the embryo was directly linked to the justification of abortion. To abort a life that has begun is a *prima facie* bad thing for anyone who gives any value to the human embryo. Depending on how much moral value is given to the embryo, abortion is either always wrong or can be condoned in circumstances where it can be justified by balancing it against even greater evil. However, the discussion on the ethics of the beginning of life was forced to refocus, when, in the mid 1970s the technologies to fertilise eggs *in vitro* became possible. Suddenly the destruction of embryos could be a part of what those who had opposed abortion considered to be of utmost value, creating new lives.

The world's first test tube baby, Louise Brown, was born in 1978. Infertility had become an increasing problem and the fact that there was now a technological solution to it was greeted with enthusiasm by many. However, the assisted reproduction techniques in which the egg is fertilised outside the woman's body came with a price. To maximise the possibility of a successful pregnancy, more embryos than will actually be used need to be created. As a result, fertility clinics store a large number of extra embryos, many of which will never be used. An ethical dilemma now faced those who had held that embryos deserve full protection from the beginning, but who also thought that creating new lives is a good thing and that being able to help those who suffered from fertility was an excellent development. Many of these people could be identified as "pro-lifers". A partial solution to the problem was to freeze the surplus embryos, since they could, in theory, be used for further pregnancies. But as everyone knows, not all the embryos will ever be used and their destruction was actually only being postponed.[7]

Some of the early pro-lifers adopted a less stringent position regarding the moral value of early embryos that allowed them to condone IVF, while some held on to their original beliefs and could not accept fertility treatments which resulted in surplus embryos. Those who took the middle ground campaigned for IVF methods that would require the smallest possible number of embryos (even if this meant a significant increase in financial costs and in physical burden to the women, and in psychological burden to all those involved).[8]

A technological breakthrough that is partly already with us, which is going to challenge the debate yet again, is the possibility of creating embryos without fertilisation. I am talking about cloning, and perhaps also of stem cells, here. The method of nuclear transfer, which created Dolly the first ever cloned mammal,[9] does not include fertilisation and could, in theory, be applied to humans as well. The possibility of creating human beings through cloning has already had an impact on, for instance, legislation in the United Kingdom where the embryo used to be defined as a result of fertilisation.[10] With the new cloning technology, however, fertilisation is not necessary for a new life to begin, so those who wish to hold on to the sanctity of human life from the very beginning need to rephrase what they mean by its beginning. A perhaps even more far reaching problem could arise if it became technologically possible to reprogram stem cells to, once again, produce all the cells of the body, because then a human being could, in theory, be created from any stem cell. It was first thought that adult stem cells could be found only in the spinal cord and in cells that take part in reproduction, but lately they have been found in various places, such as in fat cells and brain cells.

So far the most versatile stem cells (those that can be programmed to create a huge number of different types of cells) have been found in the inner cell mass of early embryos, so-called embryonic stem cells, but there is also research suggesting that adult stem cells could be programmed to generate different types of cells. It can only be guessed what sort of ethical pandemonium would follow if someone started to protect stem cells as "potential embryos".

4. End of Life

Ethical issues related to end of life have been central to medical ethics from the very beginning. The *Hippocratic Oath* from 500 BCE contains a passage that forbids doctors to give patients medicine that would hasten their death. And again, the ethical dispute on whether, for instance, suicide is ever permissible, and under what circumstances, if any, could euthanasia be justified, seems never ending.[11] In addition to these larger-than-life disagreements, technological advances have shaped, and continue to shape, these debates. The vast advances in life support technologies have made it possible to keep patients "alive" almost indefinitely. These developments have forced us both to redefine death and to reconsider the difference between "ordinary" and "extraordinary" means of keeping patients alive.

Back in the old days cessation of spontaneous breathing and heartbeat were considered to be clear signs of death. It is still true that if the human body is deprived of oxygen for a significant period of time, death will follow. However, now we have a number of resuscitation techniques with high success rates and a vast choice of life-supporting machines that can be used to replace the biological functions of the body. And in modern medicine spontaneous breathing and heartbeat are temporarily discontinued during some major surgical operations, and people who are unable to breathe for themselves are connected to ventilators, sometimes for years. As a result of these technological advances, death came to be defined in terms of brain death.

The concept of brain death is somewhat ambiguous both in medical and philosophical terms, and at least three different types of definitions have been given. These are the whole brain formulation, brain stem formulation and higher brain formulation.[12] For the purposes of this paper, suffice it to say that technology is needed in diagnosing these different states, and that what exactly counts as death is, ultimately, a philosophical question.

Interestingly, advances in another field of medicine, namely organ transplantation, contributed, together with the advances in life-support machines and resuscitation techniques, to the acceptance of the brain death definition of death. One of the crucial factors in successful organ transplantation is that the transplants have not started to deteriorate, and this begins very shortly after the oxygen supply to the organ stops. The optimal conditions for harvesting organs for transplantation occur when the people whose organs are used are still biologically "alive". The introduction of brain death as a criterion of death has made this possible. People who are, by this definition, dead can be kept breathing and their circulation going even throughout the removal of the organs if this is needed. The harvesting of fully functioning vital organs would not have been possible before death was defined as brain death.

Euthanasia has always evoked strong ethical opinions both for and against it. Those who oppose euthanasia hold that doctors should never intentionally and actively kill their patients. The introduction of highly advanced life-support methods and medical interventions that could be used to postpone death even in situations where the benefits to the patient are non-existent have led most to believe that doctors, nevertheless, do not always have the duty to do everything they possibly could to keep their patients alive. Technological advances together with scarcity of resources have led to difficult dilemmas as to what exactly the doctors have a duty to do. Sometimes a distinction has been made between ordinary and extraordinary means of keeping patients alive, but obviously what counts as extraordinary shifts over time.[13]

One the most recent discussions in the field of palliative care has been the option of terminal sedation. In terminal sedation the patient is anaesthetized and all life-supporting care (including nutrition and water) is withdrawn. This will inevitably lead to death. The death, however, will not be caused by external inputs but will take its "natural" course. The techniques required have been around for some time, but the practice has only recently become used and discussed. It has been argued that terminal sedation will accomplish all that those arguing for active euthanasia have asked for, yet it does not include the active killing of a patient, which, to many, is categorically forbidden.[14] The general anaesthesia[15] is introduced as the ultimate means of pain management, whilst there is very little controversy in withdrawing futile care (this would be the case for instance with late stages of terminal cancer). It remains to be seen whether those who have been against active euthanasia because it involves actively killing the patient will accept this approach, or whether they will they rephrase their ethical arguments so that some passive methods become forbidden, too.

5. Genetics

The double helix of DNA was discovered more than half a century ago, but the actual applications of (human) genetics are only now starting to challenge our ethical frameworks. Of the number of challenges that genetics pose on our ethical thinking, I will consider here two in more detail. They both have to do with the rights of individuals, and the available and forthcoming technologies to acquire genetic information and make use of it. The first of these relates to the nature of genetic information. The fact that genetic information is always also information about our blood relatives creates various problems to modern individualistic ethics. The second aspect that I am going to consider is the issue of consent. The vast possibilities that genetics are expected to provide in the future have been brought in to challenge the need for informed consent when, for instance, people participate in genetic research projects. This could be seen as a significant step back as far as patients' and research subjects' rights are concerned.

There are, of course, a huge number of other ways and instances in which genetics is changing our medicine-related ethical thinking. The already unclear concepts of health and disease are becoming more difficult to define as most of us are expected to carry in our genomes many mutations that put us into one health risk bracket or another. What sort of impact should this information have?[16] With the more advanced genetic tests and therapies, the distinction between enhancements and curing diseases becomes blurred. This will be especially complicated if the genetic bases for complex character traits such as intelligence and musical ability become known. If health is defined as "normal species functioning"[17] as it often is, it would seem that, for instance, improving someone's IQ to the average level would not count as enhancement but rather "curing a disease". There also are, and will be, many ethical controversies around the creation of reproductive choices and "designer babies". And further, the fact that most of these novel technologies are very expensive is going to create ethical dilemmas in terms of justice. We have already seen how the new costly methods of treatment and diagnosis have put pressures on the ability of many traditional welfare states to provide the best possible treatments for all, and future technological advances in genetics will only add to this problem.[18] And this is to mention but a few of the ethical challenges that gene technologies will bring about.

Bye, bye individual rights?

When bioethics was born in the late 1960s, one of the predominant themes was patient rights. It was thought that patients needed to be empowered to fight the paternalism of doctors. This was also a time of liberation in many other fields of life. Sexual minorities began to get recognition, religions lost some of their authoritative power, women gained increasing independence, and the individual became the centre of her own life.

This development was always criticised by paternalistic doctors, public health figures, conservatives, many religious authorities and communitarians, who all, for different reasons, felt threatened by individualism. Although the age of the individual seems to have its shortcomings, as there simply are not enough safety-nets to keep everyone afloat, it has still looked as if the return of oppressive measures would be unlikely to gain majority support. Rather unexpectedly, genetics came to unite the various critics of individualism. Now we had scientific proof that we are all connected,[19] and that when it came to people close to us, to our blood relatives, we seemed to be bound by our shared DNA.

Genetic information is always also information about one's relatives and consequently, it has been argued, this gives rise to special duties and responsibilities among family members.[20] Further claims have been made that because our genetic inheritance is to a very large degree shared by all other humans, my genetic information (information about my own genetic make-up) is not actually mine in the sense that I should necessarily be the one to decide what happens to it.

These lines of thought, made possible by the scientific and technological advances in genetics, are contrary to the idea of individual rights. Genetic information can be very sensitive in nature; it can reveal unknown family secrets, like paternity issues, and it is linked to our health status and perhaps even to our traits of character. Those who have access to genetic information about us can, in theory, know more about us than we ourselves do. It should, of course, be remembered that genetics is a new and upcoming field of study, and that its long-term impacts are still far from clear. The degree to which genes determine our health status, character and personal abilities is not known. The current understanding is that genes alone are responsible for a small minority of our features and that various environmental factors play a significant role. However, this could well change in the future. I, for one, would consider this state of unawareness quite an insufficient basis for relaxing individual rights.

The advances in genetics are challenging the individualistic ethics of our time. If I were to take a genetic test that revealed that I have a genetic mutation which predisposes me to breast cancer, that information would not only be relevant to me but also to many of my female relatives.[21] This is the basis of the argument put forward by those who think that in the genetics era we should weaken some individual rights. They argue that it is in the interests of all to know about adverse genetic predispositions and that therefore I should share my test results with the relatives in question. This is especially the case if something can be done to prevent the prevalence of the condition. That is, because most gene-based conditions are not caused by genes alone, their occurrence can be hindered by lifestyle changes. Also, for some conditions, such as many cancers, an early diagnosis enhances the prognosis considerably. If I were to tell my female relatives of my test result, they could then have themselves tested and should they turn out to be carriers of the same mutation they could have yearly mammograms to secure an early diagnosis and a better prognosis. Some have gone even further to argue that even if the condition is such that nothing can be done, people would still be better off knowing this, as this knowledge would enable them to make their life plans more realistically.[22]

Against these apparently reasonable sounding claims it could be noted that perhaps people do not want this kind of information. After all, it is a well-established ethical principle in medicine that people should be asked whether they want tests to be taken. So, one might argue that people should somehow be asked whether they want to be informed of their genetic makeup in the first place.[23] Further, information about an adverse genetic disposition is not an unambiguous blessing because the lifestyle changes needed might be drastic and costly, and the follow-up studies required might not be available, or they might be risky or painful. Then there are the issues of discrimination and other wrongful uses of genetic information. So far in most countries, for instance, insurance companies have chosen not to use genetic information, but this could change.[24] We would also need to consider the psychological burden that unexpected bad news imposes on people, especially when the bad news might turn out to be mistaken as genetics develops.

Another, more philosophical, argument for giving individuals detailed information about their own genetic make-up relies on a Kantian-type concept of autonomy, according to which we need to know all the relevant facts of our situation to be able to make autonomous decisions.[25] Against this two main arguments can be presented. First, it is questionable whether and to what degree results of genetic tests count as "facts". Genetics is a field of study which evolves and in which the interpretation of results as well as the methods of testing change over time. When the mutations BRCA1 and BRCA2 responsible for some breast cancers were first discovered they were thought to predict cancer in their bearers with a high certainty, and this led to some very drastic decisions. Only later was it discovered that without a family history, the mutation itself had very little predictive power. The second challenge against this type of argument is to question the need for all relevant information as a prerequisite for autonomous decision making. For one thing, we simply cannot know all the relevant things (and who defines what is relevant?). But it can also be argued from a Millian viewpoint that while to make autonomous decisions we do need to know that genetic information exists, we can also autonomously decide against seeking it, if we so choose.[26]

A potentially alarming yet interesting feature in this discussion is the return to biology as the basis of ethics. We have long fought the view that biology is all there is to us, and placed special value on our mental abilities and our spiritual side, but with the emphasis on genetic inheritance the clock seems to be turning back.

What is so important about consent anyway?

The fact that technologies to acquire genetic information exist and that the information is familial in nature might, regardless of the problems indicated above, force us to reconsider some of the individualistic tendencies of our ethics. Another instance in medical ethics where gene technologies are given as grounds for changing our ethical thinking is the issue of consent. I have already pointed out how the familial nature of genetic information has led to views which seem to forget the issue of consent altogether. This occurs when disclosing information to family members is discussed without stopping to think whether they would want to know, and how their consent could be secured.[27] In these discussions the emphasis is usually on the person being tested and whether she is willing to share her test results with her family members. For disclosure, harm-to-others or for-the-good-of-the-others types of arguments are usually used. Many seem to be of the view that in the interest of the greater good, an individual's unwillingness to share her genetic make-up with her relatives should be overruled. This in itself can be seen as controversial, but a more major issue, in my view, is that the rights of the relatives who will get the information out of the blue are dismissed, or they are only considered in terms of the "objective medical good". This is in stark contrast with the requirement of consent, which was introduced partly because the medical good is not the only relevant value to be considered. People are allowed to refuse medical treatments and tests, however medically beneficial, based on their own values and preferences. This is what consent is all about; it was introduced to guarantee that individual choice and autonomy are respected.

Another reason given for relaxing the requirement of informed consent in relation to genetics is "in the interests of science". In most countries there is a legal duty to obtain informed consent from research subjects,[28] but many people are now arguing that we should not demand full informed consent for participating, say, in large-scale genetic database studies. To fulfil the traditional criteria, consent needs to be given freely (without undue influence or coercion), by a person capable of giving her consent (of age and of required mental capacities), and it needs to be based on relevant information (what is being proposed, what are the risks, what are the implications etc.).[29] The problem with genetics is that as science evolves very quickly, it is very difficult to predict all the possible future research purposes for which genetic samples could be used. For these future uses, therefore, informed consent cannot be given, because a person cannot be fully informed about something which is still largely unknown. The current research protocol requires a new consent to be obtained if samples given for a particular purpose are used for another purpose. This, it has been argued, would not work with the national databanks being established in Estonia[30] and Iceland,[31] or with the large scale gene banks in the United Kingdom and elsewhere. It would be very costly and difficult to contact all the people involved, and as the U.K. Biobank,[32] for instance, is intended to serve as a research base for any number of studies, acquiring fresh permissions from all those participating could mean that people would be approached daily.

It is clear that genetic data banks would be easier to utilise if informed consent were not made a requirement for each new study. Perhaps we should, however, consider why informed consent became such an important factor in research ethics to begin with and whether the expected benefits of genetics are indeed sufficient to overrule it, and as well as whether there are other related factors to be considered.

Since the Nuremberg trials, the idea of informed consent has served as an expression of respect for individual autonomy.[33] To seek for an informed consent from a research subject is to respect the research subject's autonomy. The discussions on what type of consent should be required for genetic data bases has brought the two types of autonomy, the Kantian and the Millian, back to the centre of the debates. An interesting feature in these debates is that some of those who say that individuals should know their own genetic constitutions (even if they themselves do not want to), because only this would enable them to make truly autonomous decisions, also suggest that we should settle for less than informed consent when people donate tissue samples or data to biobanks.[34] In the former case individual choice is overruled by demands of (Kantian) autonomy and in the latter case (Kantian) autonomy is trumped by utilitarian principles and scientific optimism. Perhaps one way to proceed would be to try to find out which type of autonomy the principle of informed consent is meant to protect.

Leaving, however, issues of autonomy on one side for a moment, there is another angle from which the matter of consent can be approached. We can assume the utilitarian stance, but ask whether the expected benefits of genetics are significant enough to give us conclusive reasons to alter our ethical frameworks. Arguably, all research with human subjects would always have been easier without rules about consent, and more results would have been gained in less time and with fewer resources. Until now, however, it has been held that expected scientific benefits cannot take precedence over the entitlement individuals have to protection. What, if anything, has genetics changed or added to this equation? I would argue that the burden of proof lies upon those who wish to disregard consent. They, I believe, are the ones who should show what undeniable benefits will flow from applying more relaxed rules of consent in genetic research. But even if we viewed the possibilities of future gene technologies with great hopes, we would still also need to show that the benefits would be justly distributed. The latter requirement is, I think, frequently overlooked by science enthusiasts. Yet, what is the justification force of benefits to a few, if many are harmed, either directly or indirectly, in the process?

In addition to considerations of autonomy, utility and justice, a number of more practical issues surround a more open consent in genetic research. At least the following questions need to be answered. To what degree would the individual who consents to give a DNA sample be able to dictate the type of research for which her material may be used? How do we make sure that the criteria remain constant over time – as perceived by all those involved? Who should have access to the material and under what conditions? If the material is not completely anonymised, can anyone guarantee its safety? May genetic databases be sold? Under which circumstances can the participants be contacted by the researchers? It seems unlikely that any "scientifically satisfactory" combination of these questions would be able to respect individual autonomy in any of its traditional forms.

In terms of scientific breakthroughs, genetic research holds many promises. In terms of individual well-being, the question is more complex. We do not yet know what genetics will reveal, nor do we know who, if anyone, is going to benefit from it. We do know that genetic information is very sensitive and, even taking into account its familial aspect, very personal. But we do not know who exactly would be interested in using this sensitive and personal information, and for what purposes.

6. Conclusion

The technological advances in genetics, and the related scientific hype, have put pressure on the core values of medical ethics. In the past, technological advances have moulded ethical discussions and values, but, I would argue, these changes are significantly different from those we are now facing. This is why we should proceed very cautiously if we are thinking of allowing our ethical frameworks to change because of advances in genetics.

When life-support technologies made it possible to keep patients alive after their spontaneous heartbeat and breathing had ceased, the material circumstances of being alive changed in a way that made it necessary to redefine death. I suppose the next big thing in this field will be the simulation of brain functions, if this can ever be achieved. That will probably mean redefining death again. As for the beginning of life, those who held that human life is sacred from the beginning but were not aware of cloning, were quite justified in emphasising the importance of fertilisation, and are now equally justified in saying that the crucial transformation occurs during fertilisation or during successful nuclear transfer. Here the reason for the change is epistemic.

Gene technologies offer no such material or epistemic justifications for an ethical paradigm change. The familial nature of genetic information might have some legitimate impact, but then again, my genetic constitution is still my genetic constitution. Although I share certain features with my blood relatives, it is my DNA (and not someone else's) that needs to be tested to get a full and accurate picture of my genetic make-up. The argument that we need to relax the requirement of informed consent, because otherwise scientific developments will be hindered, is not restricted to gene technologies. All research on humans would be easier without the demand for free and considered permissions. If informed consent is important because it upholds autonomy, then the only justification for reducing its role would be to show that autonomy is now less significant. And unless we are willing to say that, there are no moral grounds for more lenient rules of consent in genetics.

Technological advances have shaped the way in which we view the world, they have changed the material conditions under which we live and they have given us some means to control even our biological existence. Technological advances are not, however, necessary, nor do they inexorably dictate our actions. We have a choice as to whether we make use of them or not, and if we do, in what way. Developments in genetics have created a hype which is almost unprecedented in human history. But it has taken us a long time to get where we are now in terms of ethics, and especially in terms of individual human rights, and if we want to stay on this path, we should not let the ambiguous promises of science lead us astray.

Notes

1. Acknowledgement: This article was produced as a part of the projects Ethical, Legal and Social Aspects of Human Genetic Databases: A European Comparison (ELSAGEN), financed between 2002–2004 by the European Community (QLG6-CT-2001-00062), and Ethical and Social Aspects of Bioinformatics (ESABI), financed between 2004–2007 by the Academy of Finland (SA 105139). My thanks are due to these institutions for their support, and to Peter Herissone-Kelly for checking my English.
2. See e.g. Kuhse H. and Singer, P., 'What is Bioethics?' in Kuhse, H and Singer, P (eds) *A Companion to Bioethics*. Oxford: Blackwell Publishing 2001, 3–11; Jonsen, A.R., *A Short History of Medical Ethics*. Oxford: Oxford University Press, 2000.
3. http://www.wma.net
4. See Kuhse and Singer 2001.
5. Häyry, M., 'Some current issues in the ethics of biomedical research and their background in the protection of the dignity and autonomy of the vulnerable' in Häyry, M., Takala, T. and Herissone-Kelly, P. (eds) *Ethics in Biomedical Research: International Perspectives*. Amsterdam and New York: Rodopi, fortcoming.
6. For important contributions see e.g. Tooley, M., *Abortion and Infanticide*. Oxford: Clarendon Press, 1983; Steinbock, B., *Life before Birth: The Moral and Legal Status of Embryos and Fetuses*. New York and Oxford: Oxford University Press; Noonan, J.T.Jr. (ed) *The Morality of Abortion: Legal and Historical Perspectives*. Cambridge, Mass: Harvard University Press, 1970; Hursthouse, R. *Beginning Lives*. Oxford: Blackwell, 1987.
7. In most countries embryos that have been frozen for five years are should be destroyed. There is, however, no medical reason to assume that the length of the time that embryos are kept frozen makes them less suitable. In 2004 the "record time" for embryos being frozen and resulting in live birth was 12 years. See, *Human Reproduction* **1** (2004), 'Twin delivery following 12 years of human embryo cryopreservation: Case report', abstract available at: http://humrep.oupjournals.org/cgi/content/abstract/19/2/328
8. See e.g. Kass, L., *Life, Liberty and the Defence of Dignity: The Challenge for Bioethics*. Encounter Books, 2002.
9. Wilmut, I., Schnieke, A.A., McWhir, J., Kind A.J. and Campbell, K.H.S., 'Viable offspring derived from fetal and adult mammalian cells', *Nature* **385** (1997): 810–813.
10. *Human Fertilization and Embryology Act* 1990.
11. See e.g. Glover, J., *Causing Death and Saving Lives*. Harmondsworth: Penguin Books, 1977; Kuhse, H., *The Sanctity of Life Doctrine in Medicine*. Oxford: Oxford University Press: 1987; Kluge, E-H.W., *The Practice of Death*. London: Yale University Press, 1975; Grisez, G. and Boyle, J.M., *Life and Death with Liberty and Justice: A Contribution to the Euthanasia Debate*. Notre Dame, Ind.: University of Notre Dame Press, 1979; Rachels, J. *The End of Life: Euthanasia and Morality*. Oxford: Oxford University Press, 1986.
12. See e.g. Lamb, D., 'Death, Medical Aspects of' in Chadwick, R. (ed) *Encyclopedia of Applied Ethics* vol 1. San Diego: Academic Press, 1988. Most jurisdictions rely on either the whole brain formulation (e.g. the United States) or the brain stem formulation (e.g. the United Kingdom).
13. See e.g. Glover 1977, 195–196.

14. Tännsjö, T., 'Terminal sedation – a compromise in the euthanasia debate?' *Bulletin of Medical Ethics* **163** (2000): 13–22.
15. The anaesthesia is intended to be terminal, but there is no medical reason why the patient cannot be woken up from it, if that becomes necessary. In palliative care units patients are sometimes kept (by their own choice) unconscious and woken up only occasionally to eat or to meet relatives.
16. See e.g. Takala, T., 'Genetic Information and our conception of ourselves as persons' in D.C. Thomasma, D.N. Weisstub and C. Hervé (eds), *Personhood and Health Care*. Dordrecht: Kluwer Academic Publishers, 2001, 91–97.
17. This is thought to be the objective definition (or the most value free). See e.g. Buchanan, A., Brock, D.W., Daniels, N. and Wikler, D., *From Chance to Choice: Genetics & Justice*. Cambridge: Cambridge University Press, 2000, ch 4.
18. See e.g. Takala, T., 'Justice for all? The Scandinavian approach' in Rhodes, R., Battin, M. and Silvers, A. (eds), *Medicine and Social Justice: Essays on the Distribution of Health Care*. New York and Oxford: Oxford University Press, 2002, 183–190.
19. "The human genome as the common heritage" as the matter is explicated in UNESCO's Universal Declaration of the Human Genome and Human Rights (1997) available at http://www.unesco.org/shs/human_rights/hrbc.htm
20. See e.g. Rhodes, R., 'Genetic links, family ties, and social bonds: rights and responsibilities in the face of genetic knowledge', *Journal of Medicine and Philosophy* **23** (1998): 10–30.
21. Some types of breast cancer affect men also, but it is still predominantly women's disease.
22. See e.g. Harris, J. and Keywood, K., 'Ignorance, information and autonomy', *Theoretical Medicine and Bioethics* **22** (2001): 415–436.
23. To arrange this in practice will, of course, be difficult. When people are approached, they will need to be told something.
24. See e.g. Takala, T. and Gylling, H., 'Who should know about our genetic makeup and why?' *Journal of Medical Ethics* **26** (2000): 171–174.
25. See e.g. Harris and Keywood 2001.
26. See e.g. Takala, T., 'Genetic ignorance and reasonable paternalism', *Theoretical Medicine and Bioethics* **22** (2001): 485-491; Häyry, M. and Takala, T., 'Genetic Information, rights, and autonomy', *Theoretical Medicine and Bioethics* **22** (2001): 403–414, Takala, T. and Häyry, M., 'Genetic ignorance, moral obligations and social duties', *The Journal of Medicine and Philosophy* **25** (2000): 107–113.
27. See e.g. Hallowell, N., Foster, C., Eeles, R., Ardern-Jones, A., Murday, V. and Watson, M., 'Balancing autonomy and responsibility: the ethics of generating and disclosing genetic information', *Journal of Medical Ethics* **29** (2003): 74–83; Rhodes 1998.
28. Goldworth, A., 'Informed consent in the genetic age', *Cambridge Quarterly of Healthcare Ethics* **8** (1999): 393–400.
29. See e.g. Holm, S. and Bennett, R., 'Genetic research on tissues stored in tissue banks', *ISUMA: Canadian Journal of Policy Research* **2** (2001): 106–112.
30. http://www.genomics.ee
31. http://www.decode.com
32. http://www.ukbiobank.ac.uk/
33. Goldworth 1999, 393.
34. See e.g. Harris and Keywood 2001; Harris, J. 'Ethical genetic research on human subjects', *Jurimetrics: The Journal of Law, Science and Policy* **40** (1999): 77–93; also Rhodes 1998.

Quality as an Ethical Issue: Technical and Economic Aspects: A Small IT Companies Case[1]

Grażyna Ożarek
Warsaw University of Technology
Faculty of Production Engineering
Warsaw, Poland

1. Introduction

The article presents the research results which show that generating and sustaining quality require ethical foundations and the problem of quality which has been considered only in its technical and organizational aspect so far needs to be broadened. This third factor which has been neglected until now is the organizational culture based on engineering ethics. An intuitive saying that *engineering ethics and employees' morale are the most important non-technical factors influencing the creation and maintenance of the technical quality of a product or service* has been confirmed by the research. It has also turned out that *quality based on ethics is highly instrumental in creating and preserving the profitability of a company.* Particular attention has been paid to the ethical dimension of quality. It has been stressed that that the result of a given activity can be viewed in the aspect of quality only when it has been undertaken for noble reasons as only then do people benefit from it and derive satisfaction.

2. Relations between Quality and Engineering and Business Ethics

The term *engineer* means genius, wisdom, fantasy. On the one side, an engineer is associated with technical wonders, and the other with reflections of the type: should reliability and economic effectiveness be the sole assessment criteria of an engineering thought? Or: is every activity good if it brings in profit and no good if it makes a loss? In the theory of utilitarianism (English "utility" = "usefulness, benefit") it is assumed that the more beneficial some activity is the more satisfaction it brings; the worse the result is the more harmful and the least satisfactory it is. Generally, we prefer useful things and the more of our income we are ready to spend on them the more satisfaction we get, on the average [Chryssides 1999, 99–100]. The instrumental expression of our consumer's preferences are the so-called indifference maps which show how usefulness varies with the level and structure of consumption [Laider 1991, 8]. Modern utalitarian theory draws on the results of economic research and tries to adapt satisfaction measurement methods in the economic sense to solving ethical problems. Assessing engineers' activity we often refer to utilitarianism and the principle of responsibility (Table 1).

Table 1
Suggested general outline of an engineer's ethical responsibilities (source: author's analysis based on Nazar 1974)

Categories of responsibility	Expected engineers' conduct
Responsibility for reliable workmanship	According to praxeological approach – cost-effectiveness, accuracy, thrift.; work discipline-reliability, diligence, work effectiveness; Exposing low quality work
Concern for customers – recipients of engineers' services	Pro-safety activity – users' life protection; Honest information about the results of products misuse
Creating good relationships within a professional group-trust	Stimulating qualifications improvement; Encouraging cooperation, solidarity; Respecting staff's dignity; Fair awards for staff's achievements
Concern for the engineer's dignity	Systematic knowledge upgrading – concern for one's own professional development, professionalism; Venturing new innovative solutions.

The production of useful things: reliable, safe, durable, i.e. of the quality expected by customers, requires not only adequate tools and organizational methods but also depends on the attitudes of the people involved in the manufacturing process.

Many companies happen to be using the same or similar technologies, same materials, however, the quality of their products is different [Wasilewski 1998, 53; Lock 2002, 30]. The result is that organization and management theory turns more and more to the concepts which explain people's behaviour in the work process. One of them is E.Shein's model of culture which includes: *behaviour* [Schlesinger 1999, 564], *behaviour justification*, and *values, ideas and premises* influencing justifications. The most important of the three are values because they give people a sense of direction and influence what they do and how they do it.

Changes of people's attitudes are long-lasting and require adequate paradigms and efforts. The patterns of the above are set by the attitudes and behaviour of engineers who are usually part of companies' management. It is the engineering-managerial staff's attitudes and values they hold that employees' attitudes depend on, that is why engineers' ethics and organizational structure based on it are so important. This may be exemplified by a company [Foster 2001, 80] whose aim was to improve its own competitiveness. The top management started from building the company's ethical culture expressed by the Ethical Charter. The use of the Charter turned out to be very effective in practice – it boosted the staff's morale, which contributed to better quality and productivity. The prestigious effect for the company was to win the M.Baldridge Quality Prize, the jury of which rates ethical business highly.

To conclude, building new culture in an organization always requires introducing changes which should be intentional, planned by top management – Table 2.

Table 2
Intentional changes in the PDCA circle
(author's analysis based on Gasparski 1988, 145–147)

PDCA circle	Intentional changes
Plan	1. man's causative involvement in formulating changes
Do	1. on the basis of established cognition (supported by scientific knowledge)
Check	1. utilitarian assessment of changes (based on the praxeological criterion) 2. ethical assessment of changes (ethicaly positive and ethically negative) 3. esthetical assessment of changes
Act	1. reality of changes (changes cannot be superficial)

3. Empirical Analysis of the Ethical Quality Basis Research Assumptions and Thesis

It was assumed in the research that:

- engineers constitute the majority of small business owners in the technical sector
- product and service quality is economically profitable for a company
- in many sectors companies have comparable technical equipment, so manufacturing products of higher technical and economic quality than competitors depends most of all on the human factor – engineering ethics and staff's morale.

The final conclusion is if quality depends on staff's morale and quality is economically profitable then ethics is highly instrumental (ceteris paribus) in the economic growth of a company.

The overall research programme included:

- analysis of the literature on the subject,
- analysis of the IT sector specificity,
- study of chosen IT companies and their customers,
- construction of a model ethical programme for a company,
- development of ethical auditing principles for IT companies,
- verification of the accepted ethical programme in a selected company.

To conduct the studies a representative group had to be chosen among the population.

The population from which research group No 1 was selected was constituted by small IT companies of different specialties. The companies and people were chosen at random. The people under study were both university and secondary school graduates, performing different functions within a company.

The research questions were set in two questionnaires. 54 people were tested with the first questionnaire, including 12 secondary school graduates and 42 university graduates (which constituted 77% of the tested). 52 people answered the questions from the second questionnaire, including 36 university graduates (69%) and 16 secondary school graduates. The questionnaires were sent to 80 companies, of which 54 answered the questions from the first questionnaire and 52 – those from the second one. The completion percentage of statistic materials was 67.5% and 65%, respectively. The relatively high return rate of questionnaires resulted from the fact that interviews were carried out after obtaining a respondent's prior consent. The studies were conducted in Warsaw in the period 2001–2002. However, the studies can be recognized as important for the whole IT sector in Poland.

Population No 2 is IT customers – the milieu of computer science teachers at secondary schools in Warsaw. The studies concerned 134 customer-teachers. They were carried out in the period 2002–2003. 200 questionnaires were sent, of which, as mentioned earlier, 134 were answered and sent back, which meant 67% completion rate of statistical material.

The next stage was to choose a company for the case study. It was assumed that such a firm should be a small Polish company in the IT sector, owned by an engineer. The selected company produces software and satisfies all the above criteria. In this particular company the developed ethical programme was verified.

In order to obtain an answer to the principal question if there is any connection between professional ethics and the quality of products and services and how strong the connection is, the correlation (regressive) model of research was used.

The next steps of research were divided as follows:

a) Studies conducted among the staff of IT companies:
   - Quality awareness and quality-oriented activity in companies.
   - Influence of professional (engineering) ethics on quality.

- Influence of professional (engineering) ethics on the company's financial condition.
- Customers' ethical code and its influence on the company's activity.

b) Studies conducted among the customers of IT companies:
- Customers' opinion on their preferences when selecting an IT company
- Professional ethics and its influence on product/service quality.

c) case study of a specific IT company
- Verification of an ethical programme in the above company.

Concluding, the studies were based on data gathered using the technique of communicating with respondents. To this aim three questionnaires were constructed containing open, closed and ranking questions. The first study – questionnaire 1 – tests the state of quality awareness among the IT staff. The second study – questionnaire 2 – also addressed to the IT staff, was entitled: *Ethics and its influence on product and service quality in the IT sector.* The third study – questionnaire 3 – was devised to gather IT customers' opinions about the influence of ethics on quality and their cooperative preferences as to company selection.

The next step was to analyze in detail the obtained data (according to the afore mentioned plan).

4. Empirical Image of IT Companies Management

4.1. Study 1 and 2 – Quality from the IT Staff Perspective

Question: *Do you know what ISO 9000 series means?* ¾ of respondents said "Yes" (Fig. 1).

Figure 1
Answers to the question: *Do you know what ISO 9000 series means?* (devised by the author)

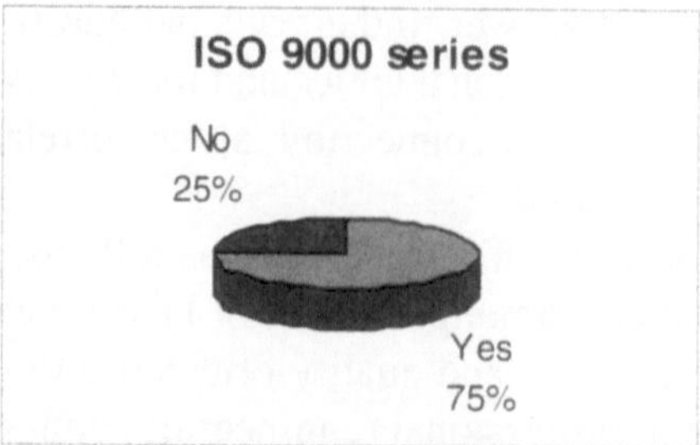

Question: *Are the problems concerning product and service quality raised in your company*?

The answers were as follows: *often* – the problem of quality is raised in less than every second company (45%); *sometimes* – in 2/5 of companies; in every seventh company quality has never been mentioned.

Figure 2
Raising the question of product and service quality in the company
(devised by author)

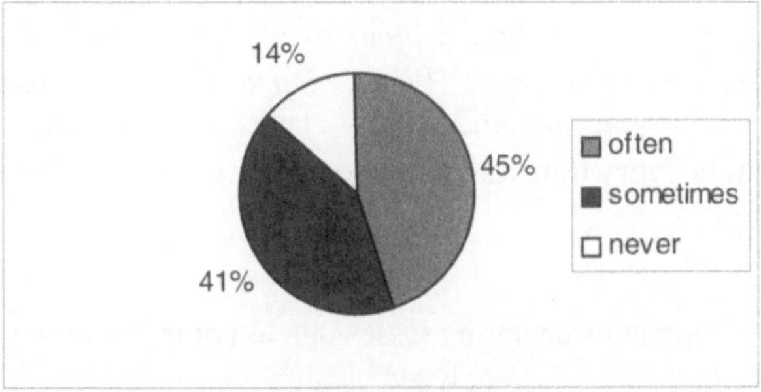

Question: *Do you see the connection between product and service quality and respecting ethical norms?* It turned out that only every eighth respondent does not see the connection between professional ethics and quality.

Figure 3
Percentage of surveyed IT companies aware of the connection between product and service quality and respecting ethical norms.
(analyzed by author)

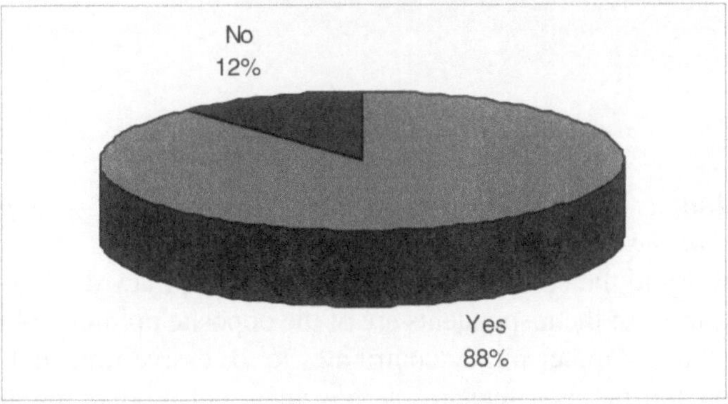

A significant majority of respondents see the strong connection between quality and professional ethics. Here are some of the exemplary replies: " it is felt intuitively, of course"; because " the worker who carries out his duties conscientiously and is fairly paid is more efficient and identifies himself with the company"; "people who are satisfied and sensibly well paid work better". However, there were some contradictory opinions – " the quality of work does not depend on ethics" – one can be a real "pro", e.g. a thief, hacker, pirate. The presented results point to the belief among the respondents that ethics is to a large extent connected with quality.

Question: *Does your company incur losses because of what is widely understood as computer piracy? If so, please give your annual loss estimate (%) and exemplary piracy type.* The obtained replies show that approximately every third surveyed company suffers losses due to computer piracy.

Figure 4
Companies declaring losses due to computer piracy
(analyzed by author)

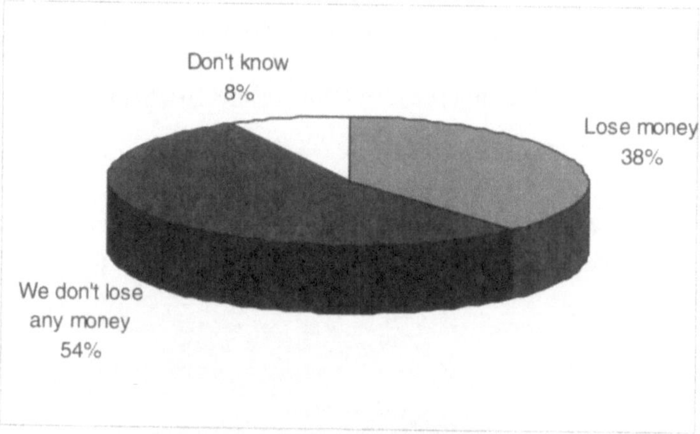

Question: *Is computer piracy in any way connected with product quality in the IT sector?*

Contrary to the common belief that computer piracy does not affect quality, most of the respondents are of the opposite opinion (64%). The fact is that computer piracy contributes to excessive time and money spent on anti-piracy safeguards at the expense of product research and development. Every fourth respondent does not perceive the role piracy in lowering the quality of IT products, saying that piracy lowers the company's profits but not the quality of its products.

Figure 5
Computer piracy and its influence on quality (analyzed by author)

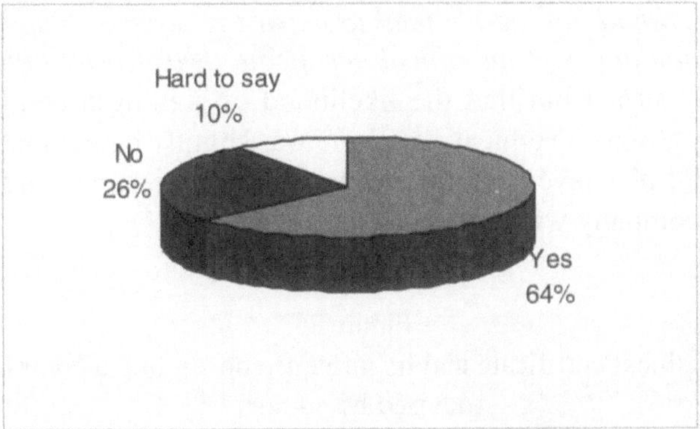

Question: *Can the conduct in keeping with professional ethics have any influence on the company's financial condition?*

Figure 6
Percentage distribution of management's opinion on the subject whether professional ethics has any influence on the company's financial condition. (devised by author)

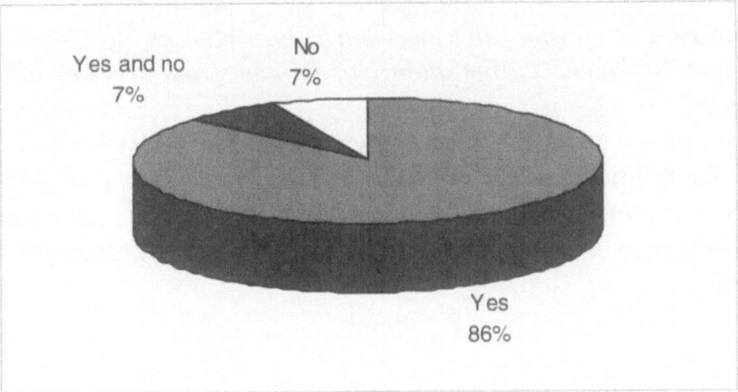

The respondents' answers show that ethical behaviour "contributes to increased sales on the barely ethical IT market"; however, they were also afraid that some people acting unethically can gain the upper hand in a very short time.

4.2 Study No 3 – Quality from the Customer's Perspective

Question: *Would you be inclined to entrust a service to an unknown company holding, e,g. an ethical certificate awarded by independent experts?* It turned out that the likelihood of signing a contract with a company having an ethical certificate was slightly higher that with the one that did not have it. Only every second customer would trust an unknown company with an ethical certificate (Fig.7).

Figure 7
An ethical certificate and its influence on signing a contract.
(devised by author)

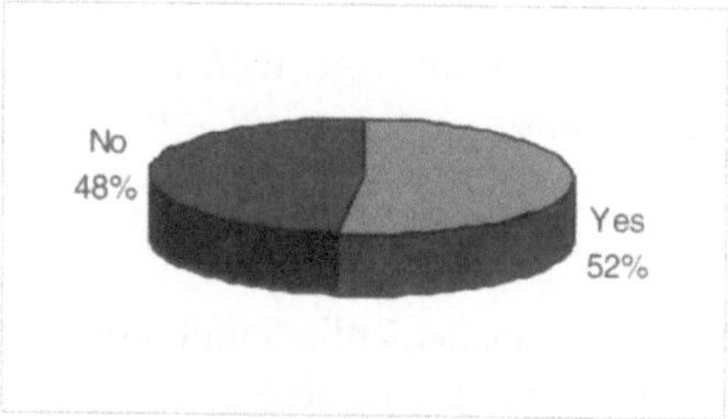

Question: *According to you, what values held by the company's staff have influence on the quality of products and services offered by them?*

The ranking results showed that values held by the staff should be put in the following order: (1) reliability and professionalism, (2) honesty, (3) kindliness towards customers, (4) promptness, according to their influence on the quality of products and services offered by the company (Fig. 8).

Figure 8
Percentage structure index showing the values held by the staff, in order of their importance, according to customers' opinion, for product and service quality
(devised by author)

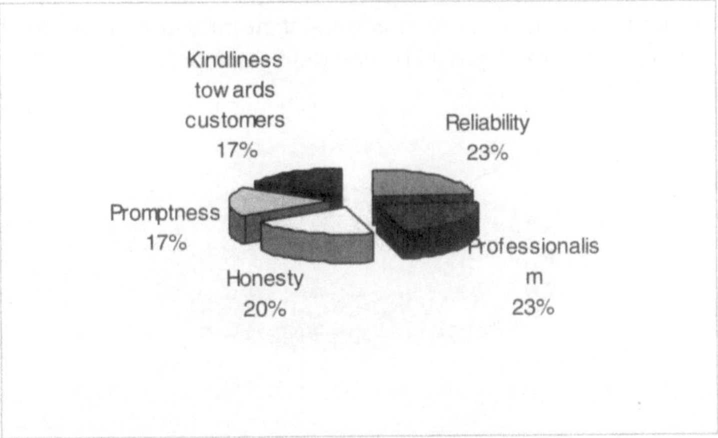

Question: *Do you see the connection between respecting professional ethical principles* and *work and service quality?*

As the replies show, the vast majority of respondents see such a connection and say that respecting ethical standards has a positive effect on the quality of work and service (Fig.9).

Figure 9
Professional ethics and its connection with service quality
(devised by author)

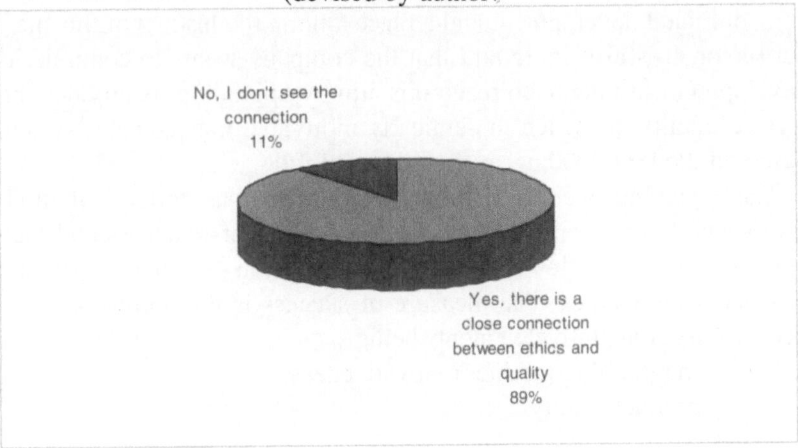

Question: *Do software piracy and hardware imitation lower the quality of IT products*? Hardware imitation and software piracy have an adverse effect on the quality of IT products. Such an opinion was expressed by IT customers (Fig.10).

Figure 10

Hardware imitation, software piracy and their influence on the quality of IT products (devised by author)

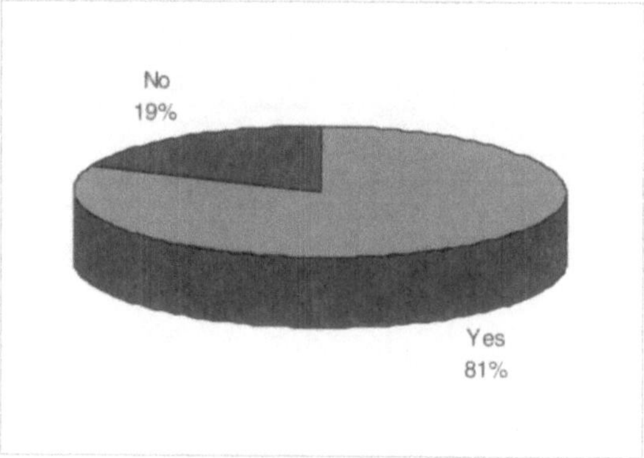

5. Application

5.1. Case Study

The analyzed company belongs to the IT sector, due to its activity. The subject of its activity is producing computer applications, IT consultancy and selling software to small and medium size companies.

The company's advantage is its almost ten-year existence on the market, continued development and a place among the leaders of this market sector. It is also important that the company wants to continue its development. It intends to reach this aim by upgrading its product and service quality, in which the ethically motivated management system based on the ISO 9000 norm should be helpful.

The beginnings were difficult as the company was started with modest means. Soon, owing to hard work it had its first success and then numerous awards followed as well as cooperation with major software producing companies. The measure of success is the number of sold licences, its constitutive elements being:

- maintaining product innovativeness,
- product quality.

Figure 11
Producers of software for small companies.
The number of sold programs in thousands until 2002,
by competitors and the company in question
(author's own source: DiS, Zwierzchowski 2003)

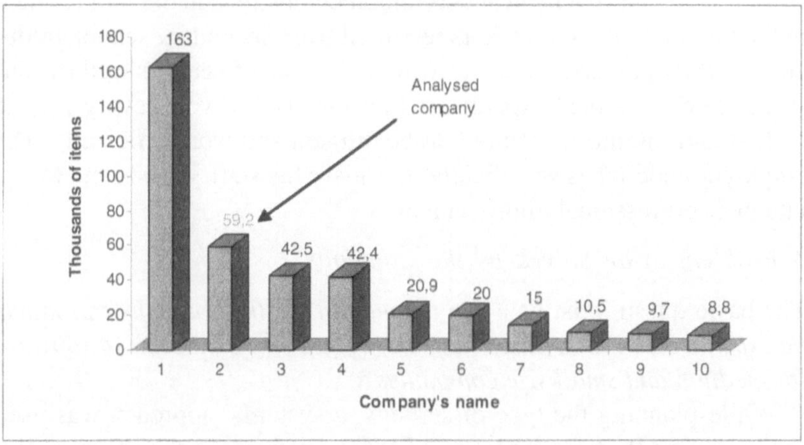

Staff's Selection

The highest requirements are imposed on programmers. The rule is that every new worker is employed on a trial basis. The company is interested in cooperating with every able and creative person, who can be offered a lot as the motivation system in the company is set to cater for higher needs, being based on the last two elements of Maslow's pyramid.

Testing Product and Service Quality

The above is dependent on the job's specificity, e.g. in the programming department it is done as follows: the work is technically analyzed as for: codes sheets, screens and lists, throughout the whole project and after its individual stages. Checking an application at its developmental stage is done by control tests. Feedback information from testers is passed on to programmers who make the necessary amendments. At the end of the programming stage a final report is made, i.e. a list is drawn up of introduced amendments or additional functions.

Corporate Culture

The company culture is usually a derivative of values held by the founder. In his dreams he built the company where family relations exist, based on friendship. This vision has been realized in real life. One can sense the atmosphere of friendship prevailing in the company. The need for constant self-improvement and development instilled in the staff resulted in the fact that the first workers recruited from secondary school graduates got their university degrees within a couple of years as well as considerable professional experience. In his everyday work, every worker is, first and foremost, required: to be diligent and work efficiently. The employer undertakes specifically: to satisfy his staff's needs and facilitate their professional improvement.

A Problem to be Solved by the Company

The basic question is: *how can the market for the company's products be expended so that it could have its biggest share in terms of software for medium and small size companies?*

While planning the type of strategy, a systemic approach was used incorporating Porter's model and SWOT analysis. In the above analysis, Porter's model was adjusted to the existing situation (in this case, the analysis of the strengths of competitors (5) was reduced to three). It concerned the analysis of: the software market sector for small and medium size companies; the position and strength of consumers; the strength of competition from other companies.

As a result, the analysis showed that stronger competition in the sector as well as inner conditions require implementing the quality system based the ISO 9000 standard, preceded by the introduction of an ethical programme which will contribute to the improvement of the existing organizational culture. Next, both the Company and the sector will have to take steps towards regulating ethical issues in order to find an answer to a question like this, for instance: *What should we do about unfair competition? How can be build an ethical infrastructure in the sector?*

5.2. The Company's Ethical Programme

The purpose of doing business is to achieve economic values by efficient activity. An ethical programme helps to improve it by all sorts of efforts aiming at the improvement of mutual trust between the parties concerned, according to the principle of mutual benefits.

The effects of implementing an ethical programme:

- Boosting the morale of the staff (lowering staff's absenteeism, lateness, unjustified downtime) and of the employers (raising staff's trust due to honest and reliable information about the company; following the principle of distributive justice, lower staff's turnover);
- higher awareness of mutual employers and staff's interests (understanding that they are mutually dependent);
- achieving a desired image of the company and reputation in the professional milieu (brand image);
- higher people's trust in the company's activity;
- solidifying and improving the company's market position;
- eliminating the company's losses caused by the wastage rate;
- raising technical and functional quality of products and services (the functional aspect of a service concerns its immaterial sphere – customer service, e.g. kindness, politeness, reliability, competence);
- reduction of transaction costs over a longer period of the company's existence;
- an increase in the company's profitability;
- minimizing problems when the company is temporarily in trouble.

Premises and required properties of an ethical programme:

- I assume (ceteris paribus) that quality depends on people, their convictions and belief that what they do is right;
- I assume (ceteris paribus) that an increase in the staff and employers' morale will contribute to mentality change in terms of their commitments;
- I assume that manufacturing products and rendering services of the quality expected by customers, not evading responsibility for the product or service, promptness and honesty, have a positive effect on the company's image, which in turn pays off in terms of customers' greater interest in the products or services of the company in the long run. The final effect of the assumed connections is the company's increased profitability.

Called-for properties of an ethical programme:

- It complies with the culture of a given environment (it adjusts the company's activity to environmental requirements).
- It has a positive effect on staff and employers' behaviour.

- It contributes to better quality, innovativeness and efficiency.
- It contributes to the development of distributive justice.
- It contributes to maintaining staff and customers' loyalty.
- It has influence on the company's profitability and contributes to its development.
- It indicates ethical standards of behaviour in morally difficult questions connected with work.
- It integrates the staff around common aims and values.
- It serves a communicative and informative function in creating a desired image of the company.
- It creates professional solidarity and need to help each other.

Elements of an ethical programme

1. The company's vision, mission.
2. The ethical code – developed in compliance with the accepted values.
3. Educational programme – devised in accordance with the mission, vision and ethical code.
4. Audit – it checks if the aims set for the system have been finally achieved.

The impact of an ethical programme on product and service quality

Table 3
Matrix of interrelationship of EP, Pvm, Qps elements
(1 – direct interrelationship, 0 – indirect interrelationship)

	EP	Pvm	Qps
EP	X	1	0
Pvm	1	X	1
Qps	0	1	X

Legend: EP – ethical programme, Pvm – people (their values and morale), Qps – quality of products and services.

The company's operation is based on two significant variables: produced quality of goods/services and supply, the result of which is profit. These variables are interconnected and logically dependent on each other. The initial parameter for the quality variable is: needs and people's requirements (the technical and economic aspect), ethical culture (inner reinforcement, e.g. distributive justice and obstacles, e.g. low morale). The output is a measurable effect – the company's profit and its image.

Table 4
Matrix of the relationship – customers' needs (input) – company's performance (output)
(devised by author)

Environment	**Customers' needs (input)**	
Company's performance – output	**Quality (technical and economic aspect)**	**Practised ethics (ethical culture of the company)**
Loss	Inadequate (design, technology, materials, price)	No (thefts, corruption)
Profit	Creativity, innovation, efficiency	Responsibility, thrift, honesty
Image (quality+ethics)= profit+safequards of the company's future income	Trust in quality, usefulness (satisfaction, prestige, pleasure)	Confidence in organizational culture, (satisfaction, prestige, pleasure)

The activity of the company is based on laws and principles of economics, philosophy, sociology and psychology. The basic problem is hidden in the question: *what number of products and services will be bought by customers from individual companies when the product (service) offered by them has similar technical parameters and price, and what differs them considerably is only ethical culture and quality?*

5.3. Ethical Programme – Implementation Procedure

Table 5
Implementation procedure IP
(devised by author on the basis of Gasparski et al. 2002, Year 2001)

Sequence	Decision on:	Activity	Resposible
Start	starting work on IP	inviting a co-worker to create and implement IP	top management-president or vice-president
1	training (training can be carried out by cascading – first top management	belief in the legitimacy of creating IP	trainer or in the cascading method-selected trained management
2	selecting the IP team	selecting team members; selecting the person in charge (IP manger) of ethical program development	president or vice-president
3	defining the aim creating an ethical code (EC)- these values should be accounted for in the wording of the EC Re: establishing common values, i.e. shared by all members of the organization, constituting its hard core.	questions should concern the following issues [Gasparski, www.rzeczpospolita.pl]: · company's mission and using the mission in advertisements and commercials; reasons why the company should have its code; · membership in other organizations which have a code (e.g. a bank, a guild); Method I: looking for values in the company's texts; developing a questionnaire. Handing out the questionnaire to the staff. Collecting the questionnaires. Analyzing the data and announcing common values. Method II: brainstorming; writing down values on the board and voting on them to make a ranking list of values.	EP leader

		Method III: dividing the staff into groups of four which make a list of common values. The values defined by teams are written down on the board looking for the common ones. Making a final ranking list of common values.	
4	defining the form of the code	e.g. concise and realistic, written in a simple and clear style, a model of reference, a shining example; or plain, legal language, etc.	EP leader
5	defining the mission	Presenting a draft version of the mission to management and staff; agreeing on a common mission	EP leader
6	defining the vision Image??	Presenting a draft version of the mission to management and staff; agreeing on a common vision.	EP leader
7	developing "a manual" of professional standards of a computer scientist	· getting to know the existing professional standards, · defining the values for the IT profession, · establishing Standards for the profession in the company.	A computer scientist, team member
7a	developing an ethical code of the company analyzing regular customers' expectations	Writing down the preamble – formulating the company's ethical policy · Formal definition of basic values · Drafting a map of the people interested. Preparing a questionnaire and interviewing regular customers. A letter explaining why the survey should be carried out, sent to the dealers · Drawing up the questionnaire;	President EP leader

	analyzing dealers' expectations	surveying the dealers Editing the code of conduct in the company Company's inner regulations: · employer-staff relationship · motivating – prizes, incentives, punishments, promotion · employee-employer relationship · ethics of decision-taking · ethics of inner and outer communication (media) · informing about unethical Conduct – formal Complaints, conflict-solving procedures Regulations concerning relations with people from the outside: · relations with business partners · relations with customers · relations with dealers Frequently asked questions and answers Defining sanction for breaking the EC	
8	implementing the code	Staff training – description of the main theses of the code, discussion, making changes or amendments – if necessary, signing the pledge to obey the established standards	president, EP leader
9	promoting the code	Putting it on the webpage or printing	president
10	programme of ethical training	**Methods of ethical training:** Seminars, workshops, e.g. on ethical decision-taking, case study; ethical e-learning **Educational content:** · 3 Es – effectiveness, economy, ethics; · source of moral standards, · ethical theories – ethics of responsibility (deontology), ethics of benefit (utilitarianism), · law and ethics; · ethical analysis, · protecting the company's confidentiality,	specialists or a competent staff member

		· ethical decision-taking, · professional ethics (ethics-quality relationship), · organization and the people concerned (ethics in relationships with people dealing with the company) · management and marketing – ethical dimension · pathologies of economic life (corruption, unfair competition) · social responsibility of the company	
11	monitoring ethical behaviour	Creating a monitoring system – assessment and reporting – checking the compliance of declarations with practice	
12	ethical audition	Preliminaries to an audition: · arranging the audit day, · defining the group of people essential for carrying out an audit, · defining the scope, methods and aims of an audit · defining the indicators (based on the mission and vision, accepted values) and ways of measurement – preparing research tools, · carrying out an audit. Analysis of obtained data · finding weak links in the behavioural chain (conduct) · developing a method to correct observed discrepancies · preparing a report – feedback information (establishing the methods of presentation and dissemination of audit results) · implementing correction methods – strengthening the links, e.g. updating an ethical code, ethical	auditing team (auditors)

		training on a given subject, monitoring a given area, etc.	
13	to be repeated in an iterative loop	Renewing an ethical programme when needed	president

6. Conclusions

The results show that the essence of any activity concerning quality should be always considered from the point of rationality, precision and ethics, in three aspects: descriptive, contrastive and axiological. It stems directly from the framework of conventionalism which says that the more good a given activity brings the more satisfaction and benefit we get, and conversely, the worse the result the more harmfulness and dissatisfaction. It is also worth stressing that a product or an activity can viewed in ethical terms only when they are undertaken for noble reasons – to achieve a high-minded purpose and when they bring satisfaction and benefit. If an activity or a product carries the stigma of evil, it can never be considered in terms of quality because it causes damage to most of the public, which raises people's objection and anger (e.g. hacking, piracy).

As it has been shown on numerous examples the quality system has to be based on ethical foundations, ethical culture, the aim of which is to introduce changes all in staff's mentality. In ethical culture, activity is base on trust and belief that every member of an organization acts fairly and with full commitment. Such a basis is required for quality systems to achieve high efficiency. It has been established that differences in the way a product is made, despite using the same machinery and materials by different organizations, are dependent on ethical culture existing in an organization. This culture is a consequence of values held by company owners, managers who are most often engineers. The significance of engineering ethics is directly proportional to an engineer's influence on the functioning of an organization, and by the same token, the quality it generates. So:

- Cultivating such values in the company as 1) reliability and professionalism, 2) honesty, 3) kindness to customers, 4) promptness – contributes to creating and maintaining quality.
- Recommending people you trust is the most important factor deciding about the selection of a cooperation partner. The com-

pany, first of all, should be perceived as honest, its additional attitudes being: a quality certificate and a renown brand name.

- High ethical culture, contrary to popular belief, contributes to economic efficiency of the company, which results from the fact that praxeological and axiological values are cultivated there.
- Cultivating ethical values established in the company would result in higher professionalism and efficiency by boosting staff's morale and improving product and service quality. It turned out that the lack of ethical refinement among engineers leads to higher running costs of the company.

Theoretical and empirical results of the research made it possible to develop the procedure of an ethical programme and to verify its practical application in a selected IT company. The company may serve a model for small organizations which plan their aims over a long period of time and intend to base their activity on the foundations of ethical quality culture. Awareness of the significance of engineering ethics for quality may contribute to a practical perception of this sphere of engineering activity.

To conclude, it should be underlined that the accepted research aims have been achieved. The research thesis based on an intuitive belief that ethical culture used in practice has a significant influence on work quality, and this, in turn,on product/service quality and consequently, on the company's cost-effectiveness, has been empirically proven.

The direction of research initialed by the present study should be continued, concerning the question of how ethically-oriented quality is generated and improved. Further research should concentrate on relations between the following variables: engineering ethics and product quality, which will enable to predict the exact value of one of the variables on the basis of the other.

Notes

1. KBN nr 2H02C10024 (2003–2004)

References

Foster S.T.: Managing Quality: An Integrative Approach. Upper Saddle River, New Jersey. Prentice Hall 2001.

Gasparski W. (ed.).: Projektoznawstwo. Elementy wiedzy o projektowaniu. (Knowledge of Design. Elements of Designing) WNT. Warszawa 1988.

Hart J.: Ethics and technology: innovation and transformation in community contexts. Cleveland Ohio. The Pilgrim Press 1997.

Hess D.: Regulating Corporate Social Performance: a new look at social accounting, auditing, and reporting. Business Ethics Quarterly. 2001 Vol. 11:2.

Kaptein M.: Auditing and developing the ethical content of organization. Kluwer Academic Publishers. Dordrecht/Boston/London 1998.

Klimczak B.: Mikroekonomia. (Microeconomics) Wroclaw. Wydawnictwo Akademii Ekonomicznej im. Oskara Langego. 1995.

Laidler D., Estrin S.: Wstep do mikroekonomii. (Introduction to Microeconomics) Warszawa. Gebethner i Ska 1991.

Nazar R.: Problemy moralne zawodu in¿yniera (Moral Problems of the Engineering Trade). PWN. Warszawa – Poznañ 1974.

Ozarek G.: Etyczne implikacje jakoœci (Ethical implications of quality). Problemy jakoœci. nr 2/2004.

Schlesinger P.F. et al: Projektowanie organizacyjne (Organizational Design). Warszawa. PWN 1999.

Wicks A.C.: The value dynamics of total quality management ethics and the foundations of TQM. Business Ethics Quarterly. Volume 11, Issue 3. 2001.

How to Achieve Ethical Goals for Business: The Rise of Technology and the Spirit of Global Corporate Responsibility

Robert Elliott Allinson
Department of Philosophy
The Chinese University of Hong Kong

The Master said, 'The superior person understands what is moral. The inferior person understands what is profitable.[1]

The Master said, 'In his dealings with the world the superior person is not invariably for or against anything. He is on the side of what is moral.[2]

Confucius said, '… the superior person … turns his thought to … what is right at the sight of gain'.[3]

The Master said, ' … a man remembers what is right at the sight of profit …'[4]

1. Introduction

My chapter is devoted to developing the position that the business person possesses an individual ethical responsibility and that the growth of technology increases both individual and the corporate ethical responsibility. An ethical vision must guide both the individual and the business institution for this to become possible. Even if one is employed by a corporation that emphasizes ethics in its mission statement, if one's own understanding of business does not include an understanding of its ethical nature, then conflicts can result between one's own understanding of business and the company's understanding. It is the duty of business schools to foster the understanding of the importance of ethics both for the individual businessperson and for business in general. A conceptual understanding of the ethical mission of business both on the individual and the corporate level is required. One cannot rely on the presence of ethical statements as part of a corporation's mission, corporate codes, ethical audits, and the work of ethical officers to accomplish this task. As technology now makes it possible for corporations to possess greater and greater impact on the world at large, the increasing power of technological production carries with it an ever increasing ethical responsibility.

2. Profit and the Ethical Agent

If one believes that the real underlying mission of business is to maximize profits, there is a danger that there will be a conflict between ethical mission statements, codes of ethics, ethical audits, ethical officers or any of the strategies a business might employ to achieve an ethical result for its business. One good way of understanding that business cannot be defined as an activity defined as making money is that such a definition is too broad as many activities share the outcome of making money including the professions of law and medicine. The activity of a physician, for example can be defined as the activity of seeking to cure disease; the activity of a lawyer can be defined as the activity of rectifying injustice. If in medicine, the primary goal was to make a profit, one would, for example, see a greater volume of patients and pay less attention to each individual patient. At an extreme, a surgeon one would recommend operations which were not needed. There is no inherent connection between healing illness and earning a fee. This is not to say that there is anything wrong with receiving payment for healing a patient. It is not as likely that one will read about a corrupt physician as that one will read about a corrupt businessman. Why is that? It is primarily because the physician does not define his profession as making money. Hence, it is difficult for him to have a conflict arise between making money and healing patients. If a businessman redefines his profession as not to be focused on making money as the primary aim, conflicts between making money and ethical goals are not as likely to arise.

One takes for granted that the physician that one sees will be an ethical agent. One even expects this. But businessmen and businesswomen are treated with suspicion. When the time comes that one can take for granted the ethical integrity of businessmen, we will know that an education in business ethics has been successful. The great Japanese ethical behavior that exists in a non-business setting offers a rich model. I can remember leaving my wallet in a train station and returning hours later to find it untouched. No one wold think of taking what was not theirs. What would it take to extend such a wonderful Japanese cultural value of honesty into the business world?

The traditional Japanese model which values employees and customers over shareholders is a model which recognizes that profit is not the ultimate goal of business. If a business is perceived of as *fundamentally* a profit making organization, then one must ultimately value those who own and would therefore profit from the business more than those who either are employed by the business or those from whom the profit is taken. If profit is the ultimate aim of the business then the employee and the customer are *tools* or *means* by which the profit is made.

Much of the literature of business ethics in the West attempts to sort out a truce between making a profit and attempting to do so within ethical limits. The difficulty of this approach is that one faces situations in which the goals of profit and ethics collide. A route to avoid situations in which profit and ethics collide is to define business in such a way as to avoid making profit its primary dimension. If one heals a patient, one hopes that one will receive a reward and if one achieves justice, one also hopes that one's efforts will be rewarded. But the proper objective of medicine is healing the ill and the proper objective of the lawyer is achieving justice. One does not place the objective of profit before the healing of the sick nor does one place the making of fees before the achievement of justice. It is important thus, to state a proper objective of business that makes it evident that its ethical dimension is primary. If medicine can be said to possess the objective of healing the sick and law can be said to possess the objective of achieving justice then business can be said to possess the objective of providing for the material well being of society. Such a short statement of the ethical objective of business can be made more full by stating that the fundamental nature of business is that it is an activity conducting with integrity that initiates, produces, communicates about and delivers quality products, employment and services that sustain the material basis of life, contribute to and improve the social welfare. When we refer to the objective of providing for the material well being of society, we can keep in mind this is but a short form for the fuller definition. What is important to understand if business is to become a proper profession, is that one will not place the making of profit before the achievement of the quality of life of the public that is served. The classic example of Johnson & Johnson pulling the Tylenol off the shelf is an illustration of how one will not risk the health of the public in order to save a loss of profit. The objective of my chapter is to offer an understanding of business activity that makes it less likely that one will feel a conflict between the goals of business and ethical considerations. One effect from this orientation is that a greater feeling of pride should ensue for the person who enters business school and the business profession.

3. Business and Responsibility

The greater the power, the greater the responsibility. Nowadays, giant corporations are sometimes more powerful than governments of nations. The influence of the multinational corporation can be immense and cross many national boundaries. The power of technology to widen the circle of production so as to make it more and more feasible that every product can find a world market makes it more necessary that a global ethics guide business production, marketing and distribution. Philosophies which were written in earlier times concerning the responsibilities of rulers as political heads of state are no longer sufficient to handle the ethical problems that face us today. Plato did not include advice for company executives because he had no idea that company executives would rise to the powerful position that they have. Plato's advice to the philosopher king is not sufficient. More than anything else today, one needs to develop advice for business leaders. The responsibility of business schools is greater even than that of law or medical schools for business takes care of the wider needs of the greater society and its actions sometimes possess greater effects for the population at large than even the actions of nation states. As technology will only increase in its power to enhance the manufacturing process, to facilitate transportation of goods and services and speed up communication, the necessity to develop an acute and a far reaching ethical concern is pressing.

4. Mission Statements

One way in which businesses have attempted to instill a sense of ethics in their corporate aims has been by articulating ethical mission statements. It is important that a company mission statement include a clear statement of one's ethical aim as the highest priority. However, such a statement must be sincerely formulated or else it is not meaningful as is noted from the case of the Snow Brands Foods Company which went bankrupt in April 2002. According to Kobayashi Kaoiri, its corporate philosophy claimed to support the health of its customers but its unsanitary practices at its Osaka's plant rendered its low-fat milk so harmful that 8,000 consumers became sick in June of 2000.[5] According to Matsuoka Toshio, Snow Brand Food Co. makes no mention of ethics of managers or employees in their mission statements.[6] If good statements are not made in good faith or are not followed by a company, then they are worthless. The ethical statement by the company must be perceived of by all in a company to be one which is truly believed in and practiced by the top management of the company. The mission statement is not enough. According to Matsuoka Toshio, Enron's mission statement included statements like 'We work with customers and prospects openly, honestly and sincerely'. At Sumitomo Corporation which experienced a loss of $2.6 billion due to illegal transactions in copper ingots by a single departmental manager, the company's mission statement stated, 'Do not engage in a business without substance for immediate profit alone.[7] What of Nomura Securities Co. whose top executives were forced to resign twice in the 1990s, forgetting their founder's maxims, 'fighting for justice and protecting honor regardless of profit'.[8]

In order to ensure that the famous Credo of Johnson & Johnson is not forgotten, the first item of which states that the first responsibility is to those who use their products and services, the company holds international Credo meetings for its managers for the purpose of getting employees to understand and put into practise these words and employees evaluate managers in terms of whether they are adhering to the mission statement. Those who are not evaluated highly enough may be replaced.[9]

But mission statements, however important, must be congruent with an understanding of the objective of business. If one enters business with the idea that one's primary goal is the maximization of profit, one may be tempted to choose opportunities which arise that offer personal profit even though these choices are in conflict with an ethical mission statement.

5. Shareholder Responsibilities

A logical extension of the ethical duties of a public company is to include the duties of its stockholder owners. Kagono Tadao stresses stockholder responsibilities over stockholder rights. He uses the interesting example of a rare painting the owner of which wanted to have buried with him in his coffin. Corporations in his analogy are like valuable paintings and are even more public in nature than valuable paintings. While the emphasis is normally on how shareholders are cheated as in the case of Enron and WorldCom, he argues that stockholders also have a responsibility not to sell out early and leave long-term losses to be dealt with by other investors.[10] With the rise of the technologies of computer trading, and their impact on the world at large, the need to develop ethics before business operations are conducted is greater now than ever before.

6. Ethics at Risk

Recent examples of accounting fraud in the USA have provided dramatic examples that unethical behavior does not always bring profit and may in fact result in loss. Alan Greenspan criticized corporate leaders for their "infectious greed" and deteriorating ethics in a speech before congress.[11] If the business of business is making money then anything goes, including creative accounting. Even now, according to Michelle Michot Foss, an economist at the University of Houston, Enron's notorious off-balance-sheet partnerships 'can be used legitimately for financing projects in high-risk countries.'[12] If business is defined as making money without regard for social responsibility, then great social harm has been produced by such an attitude. The effects of the major accounting fraud scandals of Enron and Worldcom beggar belief. Worldcom inflated its profits by $9 billion. As a result shareholders have lost some $3 billion since the news broke and 17,000 Worldcom employees have lost their jobs.[13] The company faces a possible fine of $9 billion forcing its new CEO Michael Capellas to craft a plan for the largest corporate bankruptcy in US history, involving $107 billion in assets.[14] The California public employees' retirement system – the largest state pension fund in the U.S. – is suing to regain some of the $580 million it lost in the Worldcom scandal.[15] Enron is facing 22,000 claims totaling $400 billion. Enron's (and other energy traders) profiteering from the power problems in California cost California residents $11.3 billion.[16] How do these costs compare to the $300 million that Johnson & Johnson lost when they pulled every bottle of Tylenol off their shelves worldwide?[17] According to the US State Department Annual Report on Business Practise in China, 'The Software Business Alliance estimates that more than 90% of business software used in China is pirated. They also report increasing exports of IPR-infringing goods to other countries.'[18] According to Professor Ying Chen of the Chinese Academy of Social Sciences, sham goods flood the present markets of China and the flooding of fake and inferior goods cause great harm which he details as lowering China's standard of living, harassing the operating orders of economics, hindering the development of productive forces, corrupting social values and contaminating people's spirits. For Professor Chen, cracking down on counterfeiting offenses is the most important task of commercial ethics in China.[19]

In an article on January 5, 2001 in the *Chronicle of Higher Education* by Daniel Walfish, entitled, 'Chinese Applicants to US Universities Often Resort to Shortcuts or Dishonesty,' it was reported that roughly 70% of mainland-Chinese students who attend American universities are graduates of New Oriental's test-preparation classes and last year their courses had 70,000 students enrolled in them. A Mr. Alexander and about 29 other writers are paid by the New Oriental School, a private Chinese institution to have their personal statements, resumes and letters of recommendation doctored.[20] What is most revelatory is that when they are asked if their work might mislead universities their answer is that "I hadn't thought about that" or when asked if their work might be unethical, they answer: "I haven't thought about this question". For years, according to this article, New Oriental, like other Chinese test-preparation schools, has been pirating and selling Educational Testing Service publications. Materials confiscated in 1996 and 1997 included non-public tests that could only have been stolen from examination rooms. When Chinese students boast about high scores on graduate-school admissions tests, they often cite people who can score very high, 2200 out of 2400 on the GRE without necessarily being able to speak a proper English sentence (the world average is 1600). Some Chinese students get scores on the verbal section of the exam that are wildly out of proportion to their command of English. Recent decisions by major first tier universities in America to change their admission policies to deemphasize these tests indicate that this information is correct. It shows the existence and widespread acceptance of corruption – among the young. And not only among the young – but among the intelligentsia. If corruption is considered acceptable, nay, not only acceptable but necessary and even good by the class that frequently is associated with being moral purists, then what does this say about the rest of the social classes, age groups? For here we are speaking of students, not workers. Students, not accountants.

But this example should not surprise one who is familiar with the history of China. It has long been the practice to prepare for examinations by studying the answers of others.

> Model examination essays, commentaries on the Four Books and the Classics, anthologies of prose, poetry, and other kinds of examination aids would be bought and read mostly by examinees.
>
> Students preparing for all levels of the civil service examination constituted one of the largest reading publics in the sixteenth and seventeenth centuries.[21]
>
> As was always the case, examinees were primarily interested in success at the various levels of examinations. They read examination aids to improve their chances of success.[22] Commercially produced commentaries on the Four Books were written and published primarily for helping candidates to pass examinations. It can be assumed that with perhaps a few exceptions, all candidates would read this particular type of examination aids. Their degree of familiarity with these texts might have been much greater than their knowledge of the writings of eminent thinkers such as Wang Yang-ming (1472–1529).[23]

These students in the sixteenth and the seventeenth centuries were no different from their contemporary counterparts. They knew right from wrong.

7. Does the Market System Contribute to the Problem?

The market system functions ideally for the individual producer by the control of supply. The greater the market share, the higher the degree of control the producer can exert over the supply. Ideally, one hundred per cent of the market share would define the maximum degree of control over the supply. Pricing can then be set to achieve the maximum profit without losing either to a competitive supplier or losing the demand altogether. Maximum profit can be derived from raising the price as high as possible while maintaining the greatest degree of market share. The greater the control of the market, the higher the price can be set. There is a direct proportion between control of a market share and market price. The controller can set a price that is in keeping with the goal of maximizing his or her profit to the detriment of the consumer. While it could be argued that competition which leads to a number of producers sharing a market can keep price levels down, inevitably the larger producers drive out the smaller producers and price levels are adjusted upwards. While theoretically economies of scale could encourage lower prices this is contrary to the goal of maximizing profit. Hence, the only solution is the alteration of the goal of the maximization of profit. An ethical producer need not set prices to achieve the maximum profit.

It is therefore an oversimplification to argue as the distinguished Amartya Sen with whom I find so much to agree, that "The market system works by putting a price on a commodity and the allocation between consumers is done by the intensities of the respective *willingness* to buy it at the prevailing price. When "equilibrium prices" emerge they balance demand with supply for each commodity."[24] But it is not only a matter of *willingness*. It is a matter of capability. Demand is not ethically driven. Demand is driven by the capability on the part of the purchaser of paying the price for the commodity. *Supply* is not ethically driven. The price of the supply is not necessarily ethically driven unless the producer gives up the idea of the maximization of profit. If the producer is guided by the goal of the maximization of profit as the primary aim, the price of the supply is set by the supplier in accordance with what the market – translate the financially advantaged – is capable and willing to pay. The price is thus a function of what the financially advantaged is capable and willing to pay and how high this price can be set by the supplier without losing sufficient volume of sales which would lower the profit margin. "Equilibrium prices" are at an equilibrium only for the financially advantaged. The market system, when engaged in by producers guided by the maximization of profit, whether engaged in by a state socialism or privately owned capitalism creates ethical inequities. The market system is not inherently unethical. However, if the producer is guided by the goal of profit maximization, ethical inequities are bound to occur.

8. Is Self-Interest the Source of the Ethical Problem?

It has been a commonplace to argue that self interest, sometimes accompanied with ethical concerns and sometimes not, is the source and even the honorific source of economic behavior. For example, Adam Smith, a professor of Moral Philosophy, thought that a concern for ethics and self-interest as a basis for economic behavior certainly went together. The rightfully esteemed economist Amartya Sen, for whom I have the greatest respect, has argued that, "It is important to see how Smith's whole tribute to self-interest as a motivation for exchange (best illustrated in the butcher-brewer-baker passage) can co-exist peacefully with Smith's advocacy of ethical behavior elsewhere."[25] (The famous passage reads: "It is not from the benevolence of the butcher, the brewer, or the baker, that we expect our dinner, but from their regard to their own interest. We address ourselves, not to their humanity but to their self-love, and never talk to them of our own necessities but of their advantages.")[26] And again from Sen, "The concern of the different parties with their own interests certainly can adequately *motivate* all of them to take part in the exchange from which each benefits."[27] But the problem, as I see it, is that while self-interest may be an *adequate* motivation for business behavior, is it the *best* motivation? Self-interest, however tempered it may be argued that this should be by ethical concerns, once let loose, is not so easily constrained. It is an odd thing how self-interest in many cases seems to lead to greed.

Sen argues that self-interest as a motivation for exchange does not possess any dire effects. Indeed, he argues that a motivation for exchange is separate from other economic concerns such as production and distribution. "The butcher-brewer-baker simplicity does not carry over to problems *of* production and distribution (and Smith never said that it did), nor to the problem as to how a system of exchange can flourish internationally."[28]

Regardless of how Smith viewed the matter, it seems to me, however, that if one is motivated by self-interest in the case of exchange, why would one be motivated by other concerns in the areas of production and distribution? Is it really possible to compartmentalize one's ethical concerns? If one is happy with the profit one makes from exchange, why should one limit one's profit by ethical concerns about the distribution of one's goods and services? If one is happy with one's profit from exchange, why should one limit one's profit by ethical concerns about *what* one should produce? At the very least, a tension is produced between one's self-interest and one's ethics. Self-interest may be an adequate motivation for exchange, but can we do better than this?

One recalls Smith's famous passage referring to the "invisible hand", '... every individual ... intends only his own gain, and he is in this, as in many other cases, led by an invisible hand to promote an end which was no part of his intention ... By pursuing his own interest he frequently promotes that of the society more effectually than when he really intends to promote it.'[29]

Was for Smith the invisible hand a built in ethical direction of the market system? Smith of course does not say this. But the market system, if it is an impersonal and abstract system, does not appear to be guided by an ethical pilot. And, if it is manipulated to a large extent by individuals or corporations pursuing their own gain, it also lacks an ethical pilot. For Smith, self-interest, tempered with ethical restraint, was bound to ensue in ethical outcomes. The problem with this theory is that it is precisely that, a theory. It possesses no empirical proof (indeed Smith did not suggest that it did) and it possesses no rational proof. It is metaphysics pure and simple. Despite that it has been taken by many to be a shibboleth, an unquestioned and unquestionable truth. But when one thinks about it, there is no inherent or logically necessary connection between the pursuit of self-interest and the good of society as a whole. In fact, there may be a negative correlation. The perfectly legal actions of certain property developers in Hong Kong, for example, have resulted in driving property prices way up. These same actions created such an oversupply that property values crashed, leaving multitudes of home owners with negative equity. The self-interest of the property developer was well served, but the good of society was not served.

There is but a very short step between the notion of self-interest as the motivation for exchange to the concept of the maximization of profit. On the most primitive level, self-interest can be philosophically defined as the satisfaction of the appetite of hunger. Appetites possess the quality of growing with each satisfaction. The more one satisfies the appetites, the more they grow. In addition, a new element enters the picture. One worries about the future. The growth of appetites combined with a fear of the future produces the state of greed.[30] Greed, so it seems in many cases, can never be satisfied. With the advent of greed, the concern for others weakens and business ethics has its work cut out for it. Why begin with a motivation for economic behavior that will make it difficult if not impossible for business ethics to have any effect? Would it not be far better to begin with a motivation for exchange (and production and distribution) that had the concern for others as its motivation in the first place? Would it not be better (and truer of human behavior) for there to be a unity of motivation for action? Does it make sense for self-interest and the profit motive to be the motivation for exchange and a concern for others to be the motivation for production and distribution?

Aristotle thought that the choice of the profit motive as a motivation for economic behavior was unethical. According to Robert Solomon, "Aristotle despised the financial community and, more generally, all of what we would call profit seeking. He argued that goods should be exchanged for their "real value", their costs, including a "fair wage" for those who produced them, but he then concluded, mistakenly, that any profit (that is, over and above costs) required some sort of theft (for where else could that "surplus value" come from.) ... All trade, Aristotle believed was a kind of exploitation."[31]

But if we cannot use profit or self-interest as the motivation for economic behavior, what can we use? Profit or self-interest as a motivation, as argued above is the manifestation of the following of the appetite of hunger as the core drive of the human species. The following of the appetite of hunger as the core drive of the human species is to depict the human being fundamentally as a consumer. If the following of the appetite of hunger and the description of the human being as a consumer is the source of the ethical problem, then what is needed is a new description of the human being.

Hunger as a description of the motivation for behavior characterizes humanity at the infant stage of human development. The basic need of the infant is to consume. The primary motivation for behavior in infancy and early childhood is self-interest. Maturation requires social development. This means that self-interest is modulated by a concern for others. To utilize self-interest as the primary motivation for adult economic behavior is to run the risk of producing a society which remains fixated at an infantile stage of social development. To a large degree, this is the problem of today's society. Consumerism is characterized by the production of and the provision of and thereby purchasing of goods and services that are not healthy or economically productive. Such consumerism driven by greed fuels the engine of capitalism. How can a system which is based on fear and greed lead to an ethical outcome? One might well wonder.

Would it not make more sense to model society after the motivation of the adult and not after the motivation of the child? If society is modeled after the motivation of the infant or the child, is it not likely that society will be characterized by infantilism? The infant pursues self-interest to the point of neglecting the interest of others and ultimately, in the absence of good parenting, to the point of her or his own detriment. When the infant cannot get what he or she wants, he or she throws a tantrum. Can this be likened to the outbreak of war at the level of the nation state? If nation states conduct themselves on the model of an infant wanting more and more without restraint, what is to prevent the ultimate outbreak of war?

9. How Can One Define Economic Behavior in terms Other than Self-Interest? From Maximizer of Profit to Caregiver

If consumption is the primary activity of the human infant, then the primary activity of the human adult is that of the provider and the caregiver. If the planet and its inhabitants are one's human family, then the primary motivation for economic activity will be the concern for the other. In economic terms, this concern can be best manifested by the creation and sharing of beneficial goods and services that serve the whole of mankind. In practical terms, this can translate into changing the charter of the corporation in ways which are similar to the charter of the corporation in the nineteenth century. For example, a certain proportion of the profits of the corporation would be paid out to the communities in the world that the corporation in any way harmed. While originally the charters of corporations required that part of the profits be returned to the community, it seems that a better use of part of the profits would be to go towards restituting any damages the corporation has inflicted on those that it has disadvantaged. Another change in the charter would be to limit the amount of shares that a CEO could own in a corporation. A further change would be to set an upper limit to the salary of the CEO (in Japanese corporations, CEO's make approximately eight times the amount as the lowest paid worker).

10. The Primacy of Production and Sharing over Consumption

From the standpoint of economics, production is the primary activity; consumption can only take place after production has occurred. *One must produce first in order to have something to consume*. Production is the fundamental economic activity. *With the rise of technology, factory productive capacities have risen dramatically.* The importance of production is heightened. One must recall that production is also the fundamental philosophical activity. If one sorts activities in terms of their value for the general meaning of life, it becomes apparent that the depiction of the human being as essentially a consumer is to mistake the means for the end. A penultimate end of the human being is survival. Survival takes precedence over consumption as an end-goal of life since from the standpoint of biology, one consumes in order to survive. One does not survive in order to consume. Such a truth finds its expression in the popular saying, one eats in order to live; one does not live in order to eat. If one lived in order to eat, one would be mistaking the means of life for its end. From a philosophical standpoint, that is, from a standpoint which inquired into the meaning of actions, such an inversion of reality would be pointless and futile. Hence, maximizing consumption cannot be a defining formula of human motivation. One would only need as much consumption as was required to live; there would be no point to maximize consumption indefinitely. In any case, survival itself cannot be a sufficient motivation for living. For survival only entails that life, with all of its pain and suffering, both physical and mental, be preserved. But for what end? To what purpose? With all the suffering and injustice that human beings must undergo, it cannot be that the ultimate purpose of life is to simply continue on to experience and witness birth, sickness, loneliness, meaninglessness, failure, humiliation, betrayal, loss, rejection, aging, infirmity and ultimately, solitary death. A life the end goal of which was simply survival would be totally unconscious, masochistic or absurd. Human beings are meaning craving animals. One's life can be rendered meaningful only if it can serve in some way to ameliorate the suffering that is humankind's lot. Production of means through which the alleviation of the suffering of life can be furthered is the only sufficient motivation that can sustain one throughout one's mortal career.

It can be shown that even in the most conventional formulas of economics that underneath the categorization of the human being as a consumer one can discover that the underlying economic activity of the human being is production and essentially a production of economic values, that is social goods and services. The satisfaction or happiness or fulfillment of the human being lies in the production of satisfying goods and services that contribute to the greater welfare of human beings as a whole. Technology is a wonderful servant but it is a terrible master.

11. Man as the Guardian or Trustee of the Planet

Ultimately, the model of man as the producer of goods and services that benefit all of mankind will be the only model that will be conducive to world survival. One must utilize this model to underlie existing systems and build new systems that rely upon this model for their motivational support. Rather than the image of man as the Ultimate Consumer or the Maximizer of Profit, the planet can only survive with the model of man as the Guardian or Trustee of the Planet. While all citizens possess the responsibility of being guardians or trustees of the planet, the business professional possesses the most power to achieve the greatest effect in preserving and enhancing the material well being of human beings. With this power comes great responsibilities. Business professionals must take the leadership role in both conducting business with a sense of personal integrity and to accomplish goals that are conducive to improving the welfare of human beings on this earth. There are certainly great models of business leaders from traditional Japan. One can think of Morita, Honda and Matushita as legendary examples. All were driven by ethical considerations. And interestingly enough such ethical considerations were no obstacle to their commercial success. Was it pure coincidence that the great days of the Japanese economy coincided with their times?

Notes

1. The *Analects*, IV, 16. While these quotations are from records of conversations of Confucius, they are all statements that are attributed to Confucius. When one quotes from the *Analects*, one must be careful to distinguish remarks attributed to Confucius from remarks attributed to one of his students.
2. The *Analects*, IV, 10.
3. The *Analects*, XVI, 10.
4. The *Analects*, XIV, 12. *Cf.*, Robert E. Allinson, *Management Without Ethics*, Kluwer International Series in Business Ethics, Kluwer Academic Publishers, Boston, London, Dordrecht, 2003, forthcoming. *Cf. Analects*, IV, 10., 16 (this would be the most obvious counsel against unethical business); XIV, 12; XVI, 10.
5. Kobayashi Kaori, 'Company Mission and Philosophy', http://www.jef.or/en/jti/200211 003.html
6. Matsuoka Toshio, 'Can a Corporate Mission and Philosophy Change Corporate Culture?' http://www.jef.or.jp/en/jti/200211_004.html
7. *Ibid.*
8. *Ibid.*
9. *Ibid.*
10. Kagono Tadao, 'Cultivating Conscientious Stockholders', http://www.jef.or.jp/en/jti/200303_009.html
11. *Op. cit.*
12. *Time Magazine*, December 30, 2002, p. 40.
13. *Ibid.*, p. 31.
14. *Ibid.*, p. 35.
15. *Ibid.*, p. 34.
16. *Ibid.*, p. 41.
17. *Ibid.*, p. 41.
18. The report was prepared by the US Embassy in Beijing for the fiscal year 2001 and was released by the Bureau of Economics and Business in July, 2000. *Cf.*, www.state.gov, p. 66.

19. Ying Chen, 'Cracking Down on Counterfeiting Offenses – The Most Important Task of Commercial Ethics in China,' Symposium on Business Ethics in China: Recent Developments and New Perspectives, Hong Kong Baptist University, Centre for Applied Ethics, 5–6 March 1999.
20. Daniel Walfish, 'Chinese Applicants to U.S. Universities Often Resort to Shortcuts or Dishonesty', *The Chronicle of Higher Education,* January 5, 2001. A52.
21. Susan Cherniack, 'Book Culture and Textual Transmission in Sung China', *Harvard Journal of Asiatic Studies* 54.1: 5–125, 1994. *Cf.*, Kai-wing Chow, 'Writing for Success: Printing, Examinations, and Intellectual Change in Late Ming China', *Late Imperial China* 17.1, 1996, pp. 120–157.
22. Chow, p. 33.
23. Chow, p. 31.
24. Amartya Sen, 'Does Business Ethics Make Economic Sense', Thomas Donaldson and Patricial H. Werhane (eds.), *Ethical Issues in Business, A Philosophical Approach*, 5th edition, Upper Saddle River, New Jersey, Prentice Hall: 1996, p. 17. (emphasis mine).
25. *Ibid.*, p. 14.
26. Adam Smith, *An Inquiry into the Nature and Causes of the Wealth of Nations.* Reprinted R.H. Campbell and A. S. Skinner (eds). Oxford: Clarendon Press, 1776, pp. 26–7. *Cf.*, Amartya Sen, *On Ethics & Economics*, London: Blackwell Publishers, 1990, p. 23.
27. Op. cit., p. 15 (emphasis his).
28. *Ibid.*, p. 14.
29. Adam Smith, *An Inquiry Into the Nature and Causes of the Wealth of Nations*, (ed) Edwin Cannan, New York: Modern Library, 1937, p. 423. Patricia Werhane has argued in her *Adam Smith and His Legacy for Modern Capitalism*, New York: Oxford University Press, 1991, p. viii, that even this quotation of Adam Smith's was intended to be understood as applicable only in the context of a level playing field and a well-defined framework of justice. *Cf.*, also, p. 130, *et passim*. Thus, Smith cannot be construed to be the patron of robber barons. In his book, *The Theory of Moral Sentiments*, he states that regarding the rules of justice: '... the most sacred regard is due to them.' He also states that 'Without this sacred regard to general rules, there is no man whose conduct can be much depended upon. It is this which constitutes the most essential difference between a man of principle and honor and a worthless fellow'. *Cf.*, Knud Hakonssen, (ed.), Adam Smith, *The Theory of Moral Sentiments*, Cambridge University Press: Cambridge, 2002, pp. 204, 189. Robert C. Solomon's comments on not misinterpreting Smith are also noteworthy. *Cf.*, Robert C. Solomon, *Ethics and Excellence, Cooperation and Integrity in Business*, New York and Oxford, Oxford University Press, 1992. But it is also true that Smith thought that *given a just society*, the pursuit of self interest frequently led in the direction of justice for society as a whole. Smith thus conceived of business interests conducted inside the framework of justice as not merely linked with ethical interests but as frequent causes of ethical outcomes. What is true is that Smith thought that given a just society, the pursuit of self interest inevitably led in the direction of justice for society as a whole. Smith thus conceived of business interests as not merely inevitably linked with ethical interests but as causes of ethical outcomes. There is however, a fundamental internal conflict that is well hidden. If it were the case that the pursuit of self-interest or profit was inherently linked to good outcomes, there would be no need to add on the proviso, that man must pay sacred regard to the rules of justice and that without the observation of the rules of justice, no man can be depended upon. If

the rules of justice are needed as a check and balance system for business behaviour it can only mean that business behaviour, on its own, is not ethical. If we define business behaviour as the pursuit of profit, then the naked pursuit of profit is unethical and ethical rules are required for its proper expression. This being the case, it cannot be maintained that self-interest is linked with ethical outcomes. For the ethical outcomes that ensue may not be entirely due to the part of the motivation that is self-interested; the ethical outcomes that ensue may be due to the part of the motivation that is due to following the rules of justice. The ethical outcomes could also be due to a mixture of the two motivations together. But in any case, if ethical rules are needed in addition to business activity, then business activity cannot be construed as ethical in itself. But if an activity cannot be perceived of as ethical in itself, how can one take pride in pursuing it as a profession? As much as Smith wanted to show a link between business activity and ethical outcomes, if he adds in the qualification of ethical rules, then he has as much as admitted that business activity, or self-interested activity, needs to be supplemented by ethical considerations and therefore self-interest is not to be trusted as leading to an ethical outcome.

If self-interest is not to be trusted as leading to an ethical outcome, is there another approach possible than defining business as pursuing one's self interest while following ethical rules? Would there not be an advantage in defining business activity in another way, as pursuing an ethical goal to start with and therefore as not needing an extrinsic system of ethics to be grafted upon it? If one can define business by ethical goals, it is far more likely that business, like law or medicine, can be regarded as a profession in which one can take pride and follow standards of professionalism.

30. For a complete picture of the desires, *cf.*, Robert Elliott Allinson, *Space, Time and the Ethical Foundations*, Part II, The Nature of Ethics and the Bio-Psychological Deduction of the Emotions, Aldershot, Burlington, Vermont: Ashgate Publishers, 2002.
31. Robert C. Solomon, 'Historicism, Commutarianism and Commerce,' Peter Koslowski (ed.), *Contemporary Economic Ethics*, Berlin: Springer, 1999, p. 119.

In Search of Corporate Social Responsibility Essence[1]

Anna Lewicka-Strzalecka
Institute of Philosophy and Sociology
Polish Academy of Sciences
Warsaw, Poland

1. Introduction

The dispute on corporate social responsibility (CSR) is one of the most intense disputes being conducted both in the theory and practice of economic life. In the theory this dispute means an encounter between two kinds of rationality – market rationality and protective rationality, and in consequence leads to the postulate to revise the role and the place of the economic subjects in the social structure. The dynamics of this dispute is being shaped by the changing framework of these types of rationality. Market rationality in the epoch of liberalisation of trade and free flow of capital and information has to go beyond the traditional reasoning and to resist the alterglobalistic argumentation. Protective rationality cannot limit itself to the postulate of assisting needy as its execution proved to be a support for helplessness, wastage, entitlement attitudes and led to the crisis of the welfare state.

The dispute on CSR is being conducted not only between opponents and proponents of this concept but also among various its supporters. The latter ones differ in understanding CSR because it is a polymorphous, ambiguous and vague conception. In practice, so diverse activities of firms are considered to be socially responsible that the sense of

CSR is being diluted. Moreover this sense proves to be so strongly conditioned by the legal, institutional and cultural environment that probably there is not one single model of CSR, and various solutions are good in various circumstances.

On the one hand CSR appears to be a new philosophy of management of the world economy, and a great ethical challenge for business. On the other hand, due to its ambiguity it can be used as a smokescreen for the bad practice of some companies and managers. While theoretical agreement on understanding of CSR is lacking, an attempt to reconstruct it from the particular activities of large corporations[2] is a promising idea.

2. Objections to CSR

Although the dispute on CSR has a long history, it is not closed and is still alive. The most known and expressive representative of the CSR opponents is Milton Friedman. This prominent economist puts forward arguments against CSR both of a deontological and utilitarian nature (Friedman, 1970). He claims that donating part of the company's profits to an outside cause means breaking the basic rules of the capitalist economy, namely the rule of keeping an agreement and the rule of respecting private property. It also means taking over taxation which is a government function. The managers have neither title nor competence to impose taxes. The most important utilitarian argument put forward by Friedman contends that CSR distorts the market, which ensures social well-being in most effective way.

Although Friedman's arguments were formulated more than thirty years ago they are willingly repeated by the contemporary opponents of CSR. For example, not so long ago, Martin Wolf wrote in the *Financial Times*: "The role of well run companies is to make profits not save the planet. Let them not make the error of confusing the two" (Wolf, 2001). In the opinion of this eminent journalist, business should stand in opposition to the nonsensical attempt to change its goals. This attempt is based on a false criticism of the market economy, a wrong perception of the role of multinational corporations and forces managers to take political decisions. This attempt can cause a slowdown in the reduction of the poverty in the world, it can put costly regulations of the world economy into practice and it threatens a form of global neo-corporatism, in which unaccountable power is shared between companies, activist groups, some international organisations and a few governments (Wolf, 2001).

Serious doubts are also expressed about the very notion of "the stakeholder" and the lack of a criteria to balance the benefits of all stakeholders. The definition of stakeholders as those who can affect or are affected by the organisation (Freeman, 1984) is so broad that even the people who do not buy the products of a firm can be regarded as its stakeholder because they can affect its profitability. Important questions showing substantial difficulties in applying stakeholder theory in practice were asked by Elaine Sternberg (2001). Some of them are the following: "Is everything that a stakeholder regards as beneficial to be included in the calculation?, How are stakeholders' divergent perceptions of benefit to be discerned and entered into the balance?, Are stakeholder interests all strictly equal?, Are some more important than others? If so, which are they? And when, and by how much, and why?" (Sternberg, 2001, 313, 314).

In the opinion of some critics of CSR, the theses supporting it are propagated so authoritatively that some firms undertake activities recognised as socially responsible counter to their interest. They are afraid of exposing themselves to over-zealous advocates of society and environment and to media running after sensations which arouse common interest but which are often unverified and prove to be false.

The CSR conception may pose a convenient pretext for the government retiring from its duties and offloading them on to business. By definition social responsibility was to be undertaken by firms voluntarily, yet more and more often they are forced or encouraged to certain social or ecological activities by law and by various kinds of political pressure.

Strong objections are also taken to social responsibility reporting based on the "Triple Bottom Line" accounting. Such reports are an increasingly popular practice of large corporations and try to be an objective measurement of CSR and an instrument of comparison between companies. Wayne Norman and Chris MacDonald claim "(a) that what is sound about the idea of a Triple Bottom Line is not novel, and (b) that what is novel about the idea is not sound" (Norman, MacDonald, 2004, p.243). They try to show that the components of social performance cannot be measured and aggregated in any objective way and that the rhetoric of the Triple Bottom Line paradigm is badly misleading.

3. CSR as an Instrument of Social Legitimisation of Business

In spite of the serious criticism presented above, the idea of CSR is becoming more and more popular. Awareness of the externalities of the economy arises both among the economists and in the society and the question of the extent to which the market model should take these externalities into account is being asked. It is becoming clear that the financial capital provided by the owners is not the only factor enabling the existence and activities of the company. There are other necessary factors such as: natural resources, partners of transaction, social, legal and technical infrastructure, public utilities. These costs cannot be ignored and should not be shifted to the particular stakeholders, to the environment or to society as a whole. The externalities must be internalised by the company and the pressure to do so exerted by the social, ecological and consumer organisations is harder and harder. CSR is becoming an instrument for avoiding social risk and is getting a means to legitimise business in society.

One of the most important arguments in favour of CSR orientation is the process of globalisation. Globalisation is a phenomenon which is at the centre of interest of its main players, scientists who are examining it, activists who are protesting against it and thousands of ordinary people who feel themselves to be either its beneficiaries or its victims. The main players of globalisation are the multinational corporations whose power should not be underestimated. Many of them have sales larger than GNP of some countries. For example, in 2003 the GNP of Poland amounted to 176,6 billion USD and the turnover of Wal-Mart was 256,3 billion of USD. The dominant position of business in society must involve taking responsibility for the whole. The state doesn't have the power to ease the side effect of the market economy caused by the multinational corporations as it had been the case of national ones.

In the age of globalisation multinational corporations have their subsidiaries in the countries with various legal systems, various ecological requirements, various levels of observing human rights and various degrees of power of public opinion. If corporate conduct is monitored only from a legal point of view the assessment is dependent on the time and the jurisdiction. The same conduct may be regarded as proper in one country and wrong in an other one. For example, as a result of the banning of use of non-biodegradable packets in certain country a corporation began to export its products to countries in which there is no such ban. From the point of view of the first country, its conduct is blameworthy, from the point of the second one it is acceptable. Reducing the duty of corporations to obey the law being in force in the country in which they operate leads to financial success at the expense of a deterioration of standards of labour and ecology. As Zygmunt Bauman (2000) expressively presented it, multinational corporations wander around the world to find "green pastures", exploit them and than abandon them without any scruples to find a new promising place. The abandoned place is often a dumping ground which must be cleaned by those who are tied to the place.

Hence, social responsibility of business emanates from its position and its role in the contemporary world. Magnitude, possibilities to act and the power of modern firms are incomparably greater than the traditional ones. Both large corporations and local firms have gained such a large influence on the quality and the style of life of society that they can't be seen only as the providers of products and services. Certain lifestyles, hierarchies of values and patterns of attitudes are transferred to consumers with the products and to employees with the company culture.

Globalisation understood in the way proposed by Percy Barnevick, president of ABB industrial group in his widely cited statement is blameworthy from the ethical point of view and causes more and more protests all over the world. He defined globalisation as the freedom for his "group of companies to invest where it wants when it wants, to produce what it wants, to buy and sell where it wants, and support the fewest restrictions possible coming from labour laws and social conventions" (Horowitz Rozenblum, 2003).

This opinion is often accompanied by the view that business and morality are two separate spheres and shouldn't be confused. Business only means making profit and the only norms which must be obeyed are the legal ones. For the proponents of the thesis on the amorality of business the view is characteristic that 'everything is allowed that is not forbidden by law'. Such a philosophy of management of a company results in increasingly worse and worse ethical standards of business, especially standards of labour, not only in the developing countries but also in the developed ones. This process is a kind of movement on a downward path, sometimes described as a race to the bottom. Therefore, neither the moral norms of individuals nor company norms of conduct should be reduced to legal norms because both of them tend to settle at the lowest level which has been described in sociology as *marginal morality* (Dylus, 1992).

Transnational companies style of acting cannot be changed in a democratic way, because the company's authorities are not responsible to the voters. The authorities of transnational companies are not limited by democratic mechanisms and probably they cannot be limited in any way other than by voluntarily undertaking obligations. That is why the concept of CSR seems to be a great opportunity to transform managers' attitudes and corporations' actions. CSR means doing more than companies are required to do by legislation. This conception shouldn't assume that an individual would be acting against his or her own interest or that the company will not be taking care of its profits because such assumptions would be unrealistic. However, it presumes that such striving can be limited not by the *invisible hand* but by a kind of ethical infrastructure which is the result of the integrated effort of business community, government and civil society. This infrastructure is being built in a long iterative process which enhances responsible competitiveness both at the level of particular companies and the whole economy. Thanks to this infrastructure, the good practices of some companies should be noticed and rewarded by market and then followed by other firms. However this process is full of discrepancies and traps.

4. Research Suggestions

CSR proved to be a multiform phenomenon and the theoretical dispute has not been able to define it in a satisfying way. The other way to understand CSR is to reconstruct its nature from the practice of multinational companies operating in various legal, institutional and cultural environments. Generally speaking, it means the process of inducting theory using case studies (Eisenhardt, 1999). The first step in this process is formulating the research questions. In connection with it some suggestions can be propounded.

CSR is often defined as operating of the company for the benefit of its all stakeholders. The important weakness of this definition is the lack of criteria according to which measures should be assigned to particular groups of stakeholders. In a normal everyday situation, the company cannot operate for the benefit of all stakeholders and the choice among them have to be made. In which way? According to their needs? According to their contribution to the company success? According to their power? According to their anticipated influence on the company's prosperity?

Pressure on the company to address CSR differs and depends on many things such as: country, sector of activity, size of the company, access to information, etc. Lack of a general policy of government and particular action to encourage CSR at local and national levels is noticeable, especially in the countries of Eastern and Central Europe. Some markets reward CSR and others discourage it. Some communities demand that companies take action in favour of society and the environment and others do not do it. In connection with that one can ask the following questions: How does the company operate in an environment which discourages CSR or is indifferent to it? If the company takes any responsibility in such an environment, then why? What does it count on? Does the company differentiate its CSR policy according to the country it operates in? How do companies fit their CSR policies to various environments? The best way to answer theses questions is to compare the branches of the same company operating in different countries. Answering these questions will help to identify where the origin of CSR is, in the company or in the environment.

In the period of economic transformation CSR is not often rewarded and companies aimed at maximising short-term performance happen to win. Such companies tend to internalise benefits and externalise costs. Internalising benefits may mean taking national property especially in the process of privatisation, acting on the edge of the law, taking unfair advantage from the legal gaps, using political connections to get various licences, reduced tax rates etc. These companies externalise their costs by burdening the local community or the whole society and the environment with the negative outcomes of their activity.

People in some countries know very little or nothing on CSR and do not expect too much from business. Social expectations are modest and often confined to requiring corporations only to obey rules. Sometimes people, especially in the roles of consumers even do not know their legal rights. The citizens of the postcommunist countries can hardly imagine a company taking an ecological action voluntarily. In a national survey, Poles were asked who should take action to improve the condition of the natural environment in the place of their residence. Only 10% of them pointed out the companies operating in this place, the remaining called their attention to the local and central authorities, citizens, ecological parties and organisations (CBOS Bulletin 2000). The perception gap between firm and societal representations of CSR may prove to be paradoxically small because of small societal expectations towards business. In such a situation one should ask: Is CSR proactive or reactive? Where is the origin of CSR? Do the companies try to create the stakeholder expectations for CSR or do they only respond to the existing demand? To answer these questions one needs to examine whether the companies undertake any actions to educate their stakeholders for CSR and how they do it. The real ethical challenge for large corporations is an undertaking CSR in spite of a lack of external demands in a discouraging environment.

On the other hand, the surroundings encouraging or demanding CSR also can have some disadvantages. Some companies in such surroundings tend to treat CSR only ceremonially. Codes of conduct are neatly written but not applied, some stakeholders or external causes are loudly supported but others are exploited. Enron had a long Code of Conduct and the firm assisted in many valuable social actions, Arthur Andersen was a big supporter of business ethics courses. Is it possible to distinguish true CSR from mystification?

One of the CSR characteristics is transparency of information which means that the company should disclose how well it performs with respect to all stakeholders. On the other hand the company has an obligation to keep secrets from outsiders, including competitors. How are these conflicting requirements reconciled in the everyday practice of the companies? At present, more and more companies are voluntarily preparing social responsibility reports. Which spheres of the company's activities should be presented in the report and at which level of openness? The report should meet the information expectations of various stakeholders, but sometimes these expectations are difficult to reconcile as some would like to know something that, in the opinion of the others, should be concealed. Where are the borders of reporting?

CSR is not only about donating a part of company's profit to particular stakeholders but also deciding on the form of using it, for example, a firm offers its products or services, time and work of its employees instead of a direct financial donation. Sometimes it meets with incomprehension and lack of acceptance. For example, dismissed employees think that the firm would do better giving them some money instead of spending it on outplacement, it means covering the expenses of a training program, help in establishing their own firm, or psychological aid. Help other than financial happens to be perceived as a sign of a paternalistic relationship to those who are helped. How does the company know what is best for its stakeholders? The other side effect of

CSR may be creating a dependency attitude. People soon get used to some welfare and then treat it as their right. How do the companies try to avoid such consequences?

At the end, one general remark connected with the next step of building theory of CSR should be made. Some doubts come to mind about the matched pair methodology using to select cases. This methodology suggests to contrast 'bad' and 'good' firms with respect of CSR. Meanwhile, there is no strong selection criterion of CSR neither in theory nor in practice and various rating agencies can give various rankings. To compare bad and good firms one has to know what CSR is but that is what we are looking for. Finding the firms having high and respectively low ranks in several rankings and fulfilling many other requirements connected with the size, sector, etc. would be extremely difficult. Because of this, CSR should be treated as a multidimensional concept and what one is looking for is the complex relationship between various properties of the company and various dimensions of CSR. These dimensions could not be reduced to the one measure given by the rating agency.

The questions presented above is just a first step in the process of building theory of CSR from case study research. The process is highly iterative and tightly linked to data. As Eisenhardt rightly notes "The rationale for defining the research questions is the same as it is in hypothesis-testing research. Without a research focus, it is easy to become overwhelmed by the volume of data" (Eisenhardt, 1999, p. 536).

Notes

1. This paper is an input paper in the RESPONSE Research Project: Understanding and Responding to Societal Demands on Corporate Responsibility. The project is supported by funding under the Sixth Resarch Framework Programme of the European Union.
2. Such attempt has been undertaken by the RESPONSE consortium.

References

Bauman, Z. 2000. *Globalization* (in Polish), Warszawa: PIW.

CBOS Bulletin. 2000. *The Ecological Awarness of Poles* (in Polish), Warsaw.

Dylus, A. 1992. *Marginal Morality as a Problem for the Catholic Social Teaching* (in Polish), Warsaw, Pallotinum.

Eisenhardt, K.M. 1999. *Building Theories from Case Study Research.* "Academy of Management Review", vol. 14. No 4, 532–550.

Freeman, R.E. 1984. *Strategic Management: A Stakeholder Approach*, Boston: Pitman.

Friedman, M. 1970. *The Social Responsibility of Business Is to Increase Its Profits,* "New York Times Magazine", September 13.

Norman, W., MacDonald, Ch. 2004. *Getting to the Bottom of 'Triple Bottom Line'.* "Business Ethics Quarterly" 14:2 April, 243–262.

Rozenblum, S.H. 2003. *Youth, Globalization and Progress*, 27th of November www.transcend.org.

Sternberg, E. 2001. *The Dangers of the Stakeholder Theory.* "Prakseologia" 141, 309–318.

Wolf, M. *Sleepwalking with the Enemy: Corporate social responsibility distorts the market by deflecting business from its primary role of profit generation.* "Financial Times" 16th May 2001.

Notes about the Authors

Timo Airaksinen (Ph.D. Turku University, Finland 1975) is Professor of Ethics and Social Philosophy, University of Helsinki, Finland. He is a Life Member of Clare Hall, Cambridge and Vice-President of the Philosophical Society of Finland and the International Berkeley Society. He is Managing Editor of *Hobbes Studies*. He has written *The Philosophy of the Marquis de Sade* (Routledge, 1995), *The Philosophy of H.P. Lovecraft* (Lang , 1988), and *The Ethics of Coercion and Authority* (University of Pittsburgh Press, 1988). He has also written high school and university textbooks in philosophy and non-religious moral education. Professor Airaksinen is a member of the International Advisory Board of the Praxiology series within which he co-edited Practical Philosophy and Action Theory (Transaction, 1993). He is a Life Honorary Member of the Learned Society of Praxiology.

Robert Elliott Allinson is a Professor in the Department of Philosophy of The Chinese University of Hong Kong. He is author or editor of six books and over 190 academic papers including Space, Time and the Ethical Foundations (Ashgate, 2002) and Understanding the Chinese Mind (Oxford University Press, 2000, tenth impression). He has received over 300 citations to his works exclusive of professional and web citations. He serves on the Editorial Boards and/or as Referee for eight international journals including Business Ethics Quarterly, the Journal of Business Ethics and the International Journal of Management and Decision Making. He has been a Visiting Professor or Fellow at Peking University, Fudan University, Yale University, Balliol College and St. Antony's College at Oxford University.

Gregory Z. Bedny, PhD, ScD, CPE, presently resides in North Orange, NJ and is a member of the faculty of Essex County College. He is an American citizen who emigrated from the former Soviet Union where he earned his Doctorate Degree (PhD) in Industrial Organizational Psychology from the Education University of Moscow and a Post-Doctorate (ScD) in Experimental Psychology from the Institute of General and Educational Psychology, Academy of Science of USSR. He is the author of many original scholarly articles and books.

William Chandon is a vice president of the Center for Breakthrough Thinking. The key idea, which infuses Dr. Chandon's life and work is that when people find and follow their passion, amazing things start to happen. He started his professional career as a trainer and developer, systems analysis, programmer analyst at Hughes Aircraft. He continued his career at Computer Sciences Corporation as a principal consultant, working on large consulting projects, nationally and internationally, which focused particularly on organizational change in response to technology infusion. Now he is a consultant for the Center for Breakthrough Thinking, focusing on leadership development and organizational transformation. In addition, he also teaches in the business program at St. Mary's College of California. Chandon received a B.S. from Washington State University, an M.A. from California State University in conflict management, and a Ph.D. from the Fielding Graduate University specializing in virtual organizations and the application of knowledge in practice. He is co-author with Gerald Nadler of *Ask the Right Questions: Creating the Answers that Work* (2003) and *Smart Questions: Learn to Ask the Right Questions for Powerful Results* (2004).

Wojciech W. Gasparski, Professor of Humanities, Dr. Sc., Vice Rector for Research of the Leon Kozminski Academy of Entrepreneurship and Management (LKAEM) Warsaw, Poland. Director and founder of the Business Ethics Centre a joint unit of LKAEM and the Institute of Philosophy & Sociology, Polish Academy of Sciences. Professor Emeritus of the Institute of Philosophy and Sociology, Polish Academy of Sciences and immediate past chairman of its Academic Council. Author, co-author, and editor of over three hundred publications. He teaches: human action theory (praxiology), business ethics, organisational design, methodology, and systems theory. He visited universities in other countries, extensively contributed international conferences and published several books and over three hundred articles as well as conference papers.

Topi Heikkerö (MTh, MA) is a Researcher at the Center for Social Ethics, Department of Systematic Theology, University of Helsinki. He is currently working on his dissertation, the working title of which is "The Question of Ethics in a Technological World." Apart from research work, his occupational experience includes science journalism and teaching philosophy in secondary schools.

William Herfel
With a PhD in philosophy William Herfel has been researching chaos and complexity in scientific models for twenty years. Currently, he is applying complex dynamics to theory, practice and research methodology of Chinese medicine at University of Western Sydney's Centre for Complementary Medicine Research. Before that he published papers applying complex dynamics to scientific methodology and practice as well as the interaction between technology and markets. He co-edited *Theories and models in scientific processes: Poznan Studies in the Philosophy of the Sciences and the Humanities* with I. Niiniluoto, W. Krajewski and R. Wójcicki. Before joining the University of Western Sydney he worked at the University of Newcastle in from 1991 to 2005. There he developed a wide-range of courses in the philosophy program including social studies of science, critiques of modernity, philosophy of human sexuality and non-Western philosophy. He also coordinated large interdisciplinary courses for engineering, psychology and commerce. In recent years his largest teaching responsibility had been Technology and Human Values, required of all engineering students at Newcastle. He also delivered this course at Newcastle University's Singapore engineering program from 2002 to 2005.

Waldemar Karwowski, Ph.D., D.Sc., P.E. is Professor of Industrial Engineering and Director of the Center for Industrial Ergonomics at the University of Louisville, Louisville, Kentucky, USA. He holds an M.S. (1978) in Production Engineering and Management from the Technical University of Wroclaw, Poland, and a Ph.D. (1982) in Industrial Engineering from Texas Tech University. Recently, he was awarded the D.Sc. (Dr Hab.) degree in Management Science, by the Institute for Organization and Management in Industry, Warsaw. Poland (June 2004). He also received Doctor of Science Honoris Causa degrees from the South Ukrainian State K.D. Ushynsky Pedagogical University of Odessa, Ukraine

(2004), and from the Technical University of Kosice, Slovakia (2006). He is the author or co-author of over 300 scientific publications in the areas of work system design, organization and management; macroergonomics; human-system integration, industrial ergonomics; neuro-fuzzy modeling and fuzzy systems.

Anna Lewicka-Strzalecka Ph.D. is a Professor at the Institute of Philosophy and Sociology, Polish Academy of Sciences, Warsaw, Poland. Her primary research interests include praxiology, practical philosophy, science of science and business ethics. She published several books and almost a hundred articles and conference papers. She is a co-founder and Vice-Chairperson of EBEN Poland.

Gerald Nadler is IBM Chair Emeritus in Engineering Management at the University of Southern California. Dr. Nadler previously was on the faculty at the University of Wisconsin, Madison, and at Washington University in St. Louis. He has accepted five invited visiting professorships, four times abroad. He is President of The Center for Breakthrough Thinking Inc., an international group of consulting affiliates. Clients include companies and organizations in all sectors and many countries. His industrial experience includes a vice-presidency of general operations for the 400- employee Artcraft Mfg. Co. (St. Louis and Pittsburgh). He served for 16 years on the Board of Directors, Intertherm, Inc. ($300M manufacturer, St. Louis). He is now a member of the USC Credit Union Board of Directors and a Commissioner on the Los Angeles County Quality and Productivity Commission. He served as an officer (including president) of several professional organizations, and is listed in Who's Who in America and eight other directories. He has received 25 honors and awards (including election to the National Academy of Engineering and the University of Wisconsin-Madison's Distinguished Service Award) for his research and over 225 articles and 15 books (the latest with William J. Chandon is *Smart Questions: Learn to Ask the Right Questions for Powerful Results*).

Grazyna Ozarek, Master of Engineering and Doctor of Economical Science (Prof. W.W. Gasparski was a supervisor). She got a Ph.D in Bussines Administration from Warsaw's Institute of Technology. Among her interests are matters of product's/service's quality and its' relations with ethics of engineering. Moreover she is interested in informatics. For a number of years she has been learning computer technology in Warsaw's Secondary Schools as well as leading courses for teachers, not only in this item, but also on the subject of education management. She is an author of many publications on subject of quality management and information technology. Currently she is governing Education Office in Warsaw as an associate director.

Ville Paukkonen is a PhD. student and a researcher at the University of Helsinki, Finland. He works on both philosophy of technology and history of early modern philosophy.

Alberto Guillermo Ranea (1950, La Plata, Argentina). PhD in Philosophy (1982, Universidad Nacional de La Plata, Argentina), with a dissertation on *Statics and Dynamics in Galileo's Mechanics*. At present Full-time Professor of History of Science and Technics at the Department of History of the University "Torcuato Di Tella " (Buenos Aires, Argentina). Fellow of the Alexander von Humboldt-Stiftung (1985; 1988-9; 1995). John Simon Guggenheim Fellow (1991). British Academy Professorship (2003, Birbeck College, University of London, UK). Publications (among others) on Galileo's reception by G. W. Leibniz and Thomas Hobbes, Leibniz's concept of action and the vis viva controversy, François Blondel's ballistics and the notion of a priori, Denis Papin and the relationships between theory and technical apparatus, René Descartes and experimentalism. He finished a critical edition of the complete correspondence and papers exchanged by G. W. Leibniz and Denis Papin.

Arto Siitonen, PhD from Turku University, is a Professor of Ethics and Social Philosohy at the University of Helsinki, Finland. His areas of expertise is as follows: history of modern philosophy, Kantian scholarship, and ethics. He is working now on a book: *From Pure Reason to Probable Experience. On Reichenbach's Kantianism and Criticism of Kant's Philosophy.*

Tuija Takala, PhD is Lecturer in Bioethics and Moral Philosophy at the University of Manchester, UK and Adjunct Professor in Practical Philosophy at the University of Helsinki, Finland. She is the Chair of the Editorial Board of the Cambridge Quarterly of Healthcare Ethics and editor of the Values in Bioethics Special Book Series (Rodopi). Her main research interests include theoretical and conceptual issues in bioethics.

Professor Ladislav Tondl, PhD. born 1924, is a professor at the Philosophical Faculty Charles University in Prague. He studied philosophy, sociology and economy at the Charles University in Prague. Associate Professor and from 1968 Full Professor. Director of the Institute for Theory and Methodology of Science at the Czechoslovak Academy 1968, 1969, President of Science Policy Commission UNESCO 1968-1970, Honourary Life member of the Association for the Philosophy of Science, USA, member of some learned societies abroad, of editorial boards of some international journals and book editions.

Because of political reasons dismissed of the Charles University after 1950 and after the invasion 1968 of the Academy, worked many years in the Institute of Information Theory and Automation in Prague engaged in the problems of logical semantics, scientific procedures, relationship between modern information theory and decision theory. After the Soviet invasion dismissed of the Academy and worked in the sphere of computer-aided design and computer-graphics in a design organization. After his full rehabilitation in 1989 Head of the Centre for Science Technology and Society Studies, Academy of Sciences Czech Republic and full professor of Charles University in Prague in the year 1990–2003.

His main publications are: *Scietific Procedures*, University Boston, Dordrecht, 1974; *Problems of Semantics*, Moscow, 1974, Boston, 1981; *Social Assessment of Technology*, West Boh. Univ., Plzeò 1992 (In Czech); *Science, Technology and Society*, Ed. Filosofia, Praha 1994 (In Czech); *Between Epistemology and Semiotics*, Ed. Filosofia, Praha 1996; *Dialogue*, Ed. Filosofia, Praha 1997 (In Czech, in German)); *Technological Thinking and Reasoning* (Chapters from the Philosophy of Technology), ed. Filosofia, Praha 1998 (In Czech); *Evaluation and Values* (Methodological Dimensions of Evaluation), Ed. Filosofia, Praha 1999 (In Czech). He is also a co-autor of several books. He has published numerous papers and contributions.